Heat and Mass Transfer in Drying of Porous Media

Advances in Drying Science and Technology
Series Editor: Arun S. Mujumdar McGill University, Quebec, Canada

Handbook of Industrial Drying, Fourth Edition
Arun S. Mujumdar

Advances in Heat Pump-Assisted Drying Technology
Vasile Minea

Computational Fluid Dynamics Simulation of Spray Dryers: An Engineer's Guide
Meng Wai Woo

Handbook of Drying of Vegetables and Vegetable Products
Min Zhang, Bhesh Bhandari, and Zhongxiang Fang

Intermittent and Nonstationary Drying Technologies: Principles and Applications
Azharul Karim and Chung-Lim Law

Thermal and Nonthermal Encapsulation Methods
Magdalini Krokida

Industrial Heat Pump-Assisted Wood Drying
Vasile Minea

Intelligent Control in Drying
Alex Martynenko and Andreas Bück

Drying of Biomass, Biosolids, and Coal: For Efficient Energy Supply and Environmental Benefits
Shusheng Pang, Sankar Bhattacharya, Junjie Yan

Drying and Roasting of Cocoa and Coffee
Ching Lik Hii and Flavio Meira Borem

Heat and Mass Transfer in Drying of Porous Media
Peng Xu, Agus P. Sasmito, and Arun S. Mujumdar

For more information about this series, please visit: www.crcpress.com

Heat and Mass Transfer in Drying of Porous Media

Editors

Peng Xu, Agus P. Sasmito, and Arun S. Mujumdar

CRC Press
Taylor & Francis Group
Boca Raton London New York

CRC Press is an imprint of the
Taylor & Francis Group, an **informa** business

CRC Press
Taylor & Francis Group
6000 Broken Sound Parkway NW, Suite 300
Boca Raton, FL 33487-2742

First issued in paperback 2020

ISBN-13: 978-1-138-49726-9 (hbk)
ISBN-13: 978-0-367-77939-9 (pbk)

Library of Congress Cataloging-in-Publication Data

Names: Xu, Peng, 1973- editor. | Sasmito, Agus P., editor. | Mujumdar, Arun S., editor.
Title: Heat and mass transfer in drying of porous media/[edited by] Peng Xu, Agus P. Sasmito, and Arun S. Mujumdar.
Description: Boca Raton: Taylor & Francis, a CRC title, part of the Taylor & Francis imprint, a member of the Taylor & Francis Group, the academic division of T&F Informa, plc, [2020] | Series: Advances in drying science and technology | Includes bibliographical references and index.
Identifiers: LCCN 2019010641 | ISBN 9781138497269 (hardback: acid-free paper) | ISBN 9781351019224 (ebook)
Subjects: LCSH: Drying. | Mass transfer. | Heat–Transmission.
Classification: LCC TP363 .H387 2020 | DDC 621.402/2–dc23
LC record available at https://lccn.loc.gov/2019010641

Visit the Taylor & Francis Web site at
http://www.taylorandfrancis.com

and the CRC Press Web site at
http://www.crcpress.com

Dedication

Dedicated to the warm support of our families.

Peng Xu
Hangzhou, China

Agus P. Sasmito
Montreal, Quebec, Canada

Arun S. Mujumdar
Toronto, Ontario, Canada

Contents

Preface

Drying of porous media/materials is central to drying technology commonly encountered in the production of foods, textiles, paper, wood, minerals, sludge, building materials, pharmaceuticals and biotechnological products. It is also a fundamental operation for a number of applications such as recovery of volatile hydrocarbons from underground oil and gas reservoirs, remediation of contaminated soils by vapor extraction, water management of proton exchange membrane fuel cells, and driving plant life through transpiration etc. Heat and mass transfer in porous media is important in daily experience and significant for environmental and engineering applications. It is therefore of great interest not only for scientific research but also for efficient operation of numerous industrial applications.

Drying of porous media is a process of moisture removal from wet porous materials via mechanical and thermal processes. It might involve multiple processes including vapor transport, heat transfer, liquid flow, solute transfer, menisci movement etc. Drying rates of porous media may exhibit complex dynamics reflecting internal transport mechanisms. The drying process is viewed as an interaction of heat and mass transport both inside and outside of the porous material. Therefore, the study of heat and mass transfer in drying of porous media not only helps understanding the drying kinetics, but also is beneficial to improve dryer design and operation.

Heat and mass transfer in drying of porous media are interdisciplinary fields for fluid mechanics, thermodynamics and chemical engineering. And heat, mass and momentum transfer as well as phase change can be considered from pore to representative elementary volume and even to dryer scale. Thus, heat and mass transfer in drying of porous media is a typical multiphysics and multiscale issue. With recent advances in computational and experimental techniques, many advancements have been made on heat and mass transfer mechanisms of drying porous media. However, the feasibility of different models and theory depends on complexity and operating conditions of drying process.

This book offers a comprehensive review of heat and mass transfer phenomena and mechanisms in drying of porous materials. It covers pore-scale and macro-scale models for heat and mass transfer in drying of porous media. It includes various drying technologies such as hot air drying, solar drying, osmotic dehydration, freeze drying, impingement drying, pulsed vacuum drying, microwave drying, radio-frequency drying, infrared drying, electrohydrodynamic drying etc. It also discusses the drying dynamics of fruits, coal, food, vegetables, woods, ceramic, fibrous porous material and size-distributed particle system. The multiphysics and multiscale mathematical models in these involved drying process and porous materials can help to understand heat and mass transfer in drying of porous media and improve industrial drying technology.

This book contains eight chapters. Chapter 1 provides an overview of transport phenomenon in porous media with specific application to drying and guidelines for the selection of drying models. It introduces various drying models including the diffusion model, single and multiphase flow models in porous media and the conjugate model with external flow. Chapter 2 covers topics ranging from the diffusion mechanism in drying process, various types of diffusion and diffusivity of porous media to curve fitting and modeling of diffusion models. The common drying technologies including solar drying, hot air drying, osmotic dehydration, freeze drying and hybrid drying on porous materials ranging from fruits to coal, food, vegetables, woods and ceramic are also introduced. Chapter 3 presents the classical pore-scale models in porous media, introduce the multiscale modeling method for porous media and review the applications of fractal geometry in drying of porous media. The fractal scaling laws for characterization of microstructures in porous media and pore-scale physical models for heat transfer, gas diffusion and fluid flow are illustrated in detail. Chapter 4 reviews the current status of heat and mass transfer simulation on intensification of drying technologies. The development of numerical simulation on heat and mass transfer enhancement of air-impingement drying and freeze drying as well as radio-frequency drying are reviewed and discussed. Chapter 5 focuses on jet impingement theory and impingement drying technology, and presents heat and mass transfer characteristics in steady, intermittent and pulsating impinging jet. The effects of intermittency and pulsation on heat transfer performance of impinging jet are discussed. Chapter 6 introduces wicking and drying phenomena in fibrous porous materials and presents optimal conditions for enhanced wicking or drying. It is expected that this chapter will advance the understanding of wicking and drying dynamics in fibrous systems and facilitate the design of engineered fibrous architectures. Chapter 7 offers a comprehensive review of the capillary valve effect on drying of porous media. Pore network models for two-phase transport in porous media with capillary valve effect have been developed and presented. Chapter 8 introduces heat and mass transfer of size-distributed particle systems as well as the relevant numerical methods under drying condition. The development of the population balance modeling and updated mathematical methods are presented, and the Taylor-series expansion method of moments for particle systems during drying processes is focused on.

This book offers useful information for researchers and students as well as engineers in drying technology, food processes, applied energy, mechanical and chemical engineering. It attempts to provide guidelines for mathematical modeling and design as well as optimization of drying of porous material.

The editors thank all of the contributors and reviewers for their great efforts in preparing this book. The support from CRC Press and great efforts of Allison Shatkin, Camilla Michae, and Teresita Munoz are also acknowledged. Financial support to Prof. Peng Xu from the following sponsors is gratefully acknowledged: National Natural Science Foundation of China (No. 51876196) and Zhejiang Provincial Natural Science Foundation of China (No. LR19E060001).

Peng Xu
China Jiliang University, China

Agus P. Sasmito
McGill University, Canada

Arun S. Mujumdar
McGill University & Western University, Canada

Contributors

Asni, Tommy
Heriot-Watt University
Malaysia Campus, Precinct 5
Putrajaya, Malaysia

Chaedir, Benitta A.
McGill University
Montreal, Quebec, Canada

Chong, Chien Hwa
Heriot-Watt University
Malaysia Campus, Precinct 5
Putrajaya, Malaysia

Fan, Jintu
Cornell University
Ithaca, New York, United States

Figiel, Adam
Wroclaw University of Environmental
and Life Sciences, 51-630
Wroclaw, Poland

Kharaghani, A.
Otto von Guericke University
Magdeburg, Sachsen-Anhalt
Germany

Kurnia, Jundika C.
Universiti Teknologi PETRONAS
Bandar Seri Iskandar, Perak Darul
Ridzuan, Malaysia

Law, Chung Lim
University of Nottingham Malaysia
Campus, Semenyih, Selangor Darul
Ehsan, Malaysia

Liu, Yueyan
China Jiliang University
Hangzhou, Zhejiang, China

Mujumdar, Arun S.
McGill University
and
Western University
Montreal, Quebec, Canada

Sasmito, Agus P.
McGill University
Montreal, Quebec, Canada

Shou, Dahua
The Hong Kong Polytechnic University
Hung Hom, Kowloon, Hong Kong

Tsotsas, E.
Otto von Guericke University
Magdeburg, Sachsen-Anhalt, Germany

Wu, R.
Shanghai Jiao Tong University
Shanghai, China

Xiao, Hong-Wei
China Agricultural University
Beijing, China

Xu, Peng
China Jiliang University
Hangzhou Zhejiang, China

Yu, Boming
Huazhong University of Science and
Technology
Wuhan, Hubei, China

Yu, Mingzhou
China Jiliang University
Hangzhou, Zhejiang, China

Yu, Xian-Long
China Agricultural University
Beijing, China

Zhao, C.Y.
Shanghai Jiao Tong University
Shanghai, China

Editors

Peng Xu is professor at China Jiliang University, China. He serves as the head of the Applied Physics Department. He earned his PhD from Huazhong University of Science and Technology and National University of Singapore under the supervision of Professor Boming Yu and Arun S. Mujumdar in 2009. He was also visiting professor of the Department of Mining and Materials Engineering at McGill University from June to December in 2015. Research interests include heat and mass transfer in porous media, multiscale and multiphysics modeling and fractal geometry and its applications in engineering. He serves as the editorial board for *Fractals* (World Scientific). Prof. Xu has published two books and more than 60 journal papers.

Agus P. Sasmito is assistant professor at the Department of Mining and Materials Engineering at McGill University, Canada. He received his PhD from the Mechanical Engineering Department at the National University of Singapore. Prof. Sasmito serves as an assistant editor for *Drying Technology*, editorial member for *CIM Journal*, professional member for SME, CIM and TPR. He is the team leader of Mine Multiphysics, and current research interests include mine ventilation, energy systems (conventional and renewable/alternative power sources), artificial ground freezing, slurry transport, minerals processing, materials handling, industrial transport processes and thermal fluid sciences and engineering. Publications include 70 journal papers, 64 conference papers, three books, seven book chapters and three technical reports.

Arun S. Mujumdar is adjunct professor at McGill University and Western University, Canada. He earned his BChemEng with distinction from the Institute of Chemical Technology, Mumbai, India, and his MEng and PhD from McGill University. He was professor of Chemical Engineering at McGill University and professor of Mechanical Engineering at the National University of Singapore. He is the founder/program chairman of the International Drying Symposium (IDS) series. His research has been featured in over 500 refereed publications. He has authored two books and edited over

60 books including the acclaimed *Handbook of Industrial Drying*. He is the editor-in-chief of *Drying Technology—An International Journal*. He is the recipient of numerous international awards including Doctor Honoris Causa from Lodz Technical University, Poland, and the University of Lyon 1, France. In 2014, he received the prestigious National Award for International Cooperation in Science and Technology as well as the esteemed Friendship Award from the Government of China.

1 Heat and Mass Transfer Phenomena in Porous Media – Application to Drying

Benitta A. Chaedir
McGill University, Montreal, Quebec, Canada

Agus P. Sasmito
McGill University, Montreal, Quebec, Canada

Arun S. Mujumdar
McGill University & Western University,
Montreal, Quebec, Canada

Peng Xu
China Jiliang University, Hangzhou, Zhejiang, China

CONTENTS

1.1   INTRODUCTION

Drying of porous materials is a common yet crucial process in many industrial sectors ranging from processing of foods, pharmaceuticals, paper, wood, ceramic and biomass to electronic packaging. Many researchers have tried to simulate and study the drying process through a variety of mathematical models. As the digital age surges, many advancements have been made to mathematical modeling in the field of drying that allow developments in scientific understanding of the phenomena. Such models can be used to explain the underlying physics associated with the heat and mass transport phenomena, along with physiochemical transformations of the products (Turner and Mujumdar 1996). Mathematical modeling and numerical simulation enable researchers to predict system behavior in order to design or improve existing experiments so that expensive and repetitive experiments can be avoided. Anomalous behavior can be analyzed by comparing it to the model-predicted behavior. Simulation results can act as a tool to further develop efficient and optimized drying techniques.

As expected, there is no single model that can represent all drying processes well for a given purpose. The feasibility of different models depends on the complexity and operating conditions of the drying process. Thus, it is essential to choose an appropriate model that represents the process as close to the real configuration as possible. No author has yet provided a methodology for selection of mathematical models of a given drying process involving porous media. The aim of this chapter is to serve as a guideline for selection of drying models.

1.2   DRYING MODELS

The degree of complexity required in a model depends on what is being described. Various models of moisture migration during drying of porous media have been proposed by many researchers. The models range from simple ones, like the diffusion model, to complex ones, like the multi-fluid model, which are outlined in the following sections.

1.2.1   Liquid Diffusion Model

One of the most-often-applied models to describe and quantify drying is the diffusion model. In a system not at equilibrium, the system is spontaneously brought to uniformity by molecular diffusion, where movement of moisture from a place of high concentration to one of low occurs. The rate at which diffusion takes place depends on the concentration gradient. (Treybal 1968) Diffusion models make it possible not only to simulate the drying kinetics but also to predict moisture and temperature distribution within the products. The knowledge can be useful for the optimization of the drying equipment and process to keep moisture and temperature uniform and for stress analysis to ensure the quality of products.

Lewis and Sherwood are known as pioneers in developing mathematical drying models by applying Fourier's law of heat conduction to drying of solids, in which the temperature and thermal diffusivity were replaced by moisture and moisture diffusivity, respectively. The diffusion model assumes that moisture transport is due only to the concentration gradient between the interior of a porous body and its surface, and that liquid evaporates at the outer boundary (albeit later the liquid diffusion model was modified to include evaporation inside the porous media substrate). The following liquid diffusion equation was used by Sherwood (1929) to calculate the moisture distribution in a solid during drying:

$$dX \, / \, dt = D_{eff} \nabla^2 X \tag{1.1}$$

where X is the moisture content and D_{eff} is the effective diffusion coefficient. The diffusion model for solids is therefore an empirical model assuming flux is proportional to the local concentration gradient where, instead of a polynomial or other algebraic equations, a differential equation is used. In drying, moisture content is generally presented in either dry or wet basis, defined respectively as

$$X = \frac{mass \ of \ water}{mass \ of \ dry \ product} = \frac{\rho_l}{\rho_s} \tag{1.2}$$

$$W = \frac{mass \ of \ water}{mass \ of \ wet \ product} = \frac{\rho_l}{\rho_s + \rho_l} = \frac{\rho_l}{\rho_b} = \frac{X}{1+X} \tag{1.3}$$

where ρ_l and ρ_b represent density of liquid (water) and density of the drying substrate, respectively. The moisture content and water concentration (c_{liq}) inside the material can be related by substituting density of drying substrate into the definition of wet basis moisture content (Kurnia et al. 2013):

$$\left(S_b - \frac{\rho_b}{M_{liq}c_{liq}} \right) X^2 + \left(S_b + 1 - \frac{\rho_b}{M_{liq}c_{liq}} \right) X + 1 = 0 \tag{1.4}$$

where S_b is the bulk shrinkage coefficient that differs for different materials and must be determined experimentally. Solving the equation analytically for X and rejecting the wrong root gives the solution for moisture content:

$$X = \frac{-b - \sqrt{b^2 - 4ac}}{2a} \tag{1.5}$$

where $a = \left(S_b - \dfrac{\rho_b}{M_{liq}c_{liq}} \right)$, $b = \left(S_b + 1 - \dfrac{\rho_b}{M_{liq}c_{liq}} \right)$, and $c = 1$.

The initial moisture content is assumed to be saturated ($c = c_0 = 1$) and the moisture flux at the boundary is a function of mass transfer coefficient (h) and moisture

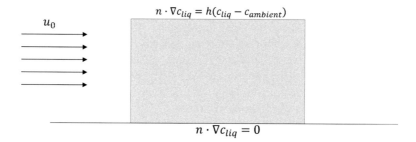

FIGURE 1.1 Schematic of the computational domain and boundary conditions for the diffusion model of drying of a rectangular slab.

concentration of the ambient (ambient humidity). In general, convection may improve the rate of moisture removal at the boundary due to a higher mass transfer coefficient. Figure 1.1 shows a schematic of a typical computational domain and boundary conditions in Cartesian coordinates.

In some cases, for simplicity, it is common to describe the complete mass transfer in drying using the diffusion equation and to take the correction for all secondary types of mass transfer into account simply by replacing the diffusion coefficient with an effective diffusion coefficient. The effective or apparent diffusion coefficient D_{eff} has no physical meaning and is determined experimentally. Generally, it is considered as a function of moisture content, temperature, material type and behavior.

The assumption of constant shape and volume of the product during drying is frequently made if shrinkage is negligible (Franco et al. 2016; Sun et al. 2005; Fan et al. 2003; Simal et al. 1996;). This assumption is often not satisfactory in modeling drying of porous media. In food and agricultural products, the shrinkage is so significant that it should not be discarded and additional information about the product is needed to model the shrinkage. However, when shrinkage is taken into account, its effect on the diffusivity should also be considered. The diffusivity obtained from empirical data assuming no shrinkage cannot be used to model shrinking materials as liquid diffusivity used in modeling drying is not a material property but an empirical parameter. Rosselló et al. (1997), Park (1998), and Resende et al. (2007) among many others have compared the results of the liquid diffusion model when considering as well as neglecting material shrinkage. Da Silva et al. (2012) compared diffusion models used to describe seedless grape drying at low temperature. The models incorporated equilibrium (first kind) and convective (third kind) boundary conditions with constant and variable volume and diffusivity. The analysis showed that a slightly higher effective mass diffusivity than the value at the final part of drying was obtained when including the effect of shrinkage but maintaining diffusivity as constant, suggesting that the change of internal structure caused by shrinkage also affects the effective mass diffusivity. From these works it has become evident that incorporating shrinkage into the drying model is necessary, especially when dealing with agricultural and food products and building materials where shrinkage is a key aspect of the dehydration phenomena. (Katekawa and Silva 2006) This conclusion leads to numerous works on drying models where shrinkage is incorporated.

For example, Mihoubi et al. (2004) developed a mathematical unidirectional and dynamic model describing the transfer phenomena (heat, mass and momentum) and deformation during drying of a thin clay sample. To account for the shrinkage behavior, the principal equations of the model are written in Lagrangian formulation and solved by a finite difference method. López-Méndez et al. (2018) studied the shrinkage-deformation behavior during drying of gel model systems (GMS) using image analysis technique. Characteristics of product deformation at different moisture contents were obtained by combining product contours to produce a shape. The obtained results demonstrate that although shrinkage occurs from the beginning of drying, changes in product shape are noticeable only after the free moisture fraction is below 0.3. Da Silva et al. (2014) used a liquid diffusion model to describe drying of whole bananas considering shrinkage and the real geometry of the product, obtained by digitization of the photography of the fruit. The effective diffusivity used was related to both drying air temperature and local dimensionless moisture content through an Arrhenius-like equation.

A list of diffusion coefficient relations for selected materials is shown in Table 1.1. Most authors assume effective diffusion coefficient to vary only as a

TABLE 1.1
Diffusion Coefficient Relations for Selected Materials

	Effective Diffusion Coefficient	Materials	References
1.	$D_{eff} = \exp\left(D_0 + \dfrac{E_a}{R(T+273)}\right)$	Green peas	Simal et al. (1996)
2.	$D_{eff} = D_0 \exp\left(-\dfrac{E_a}{R_g(T+273.15)}\right)$	Cocoa	Hii et al. (2009)
3.	$D_{eff} = D_0 \exp\left(-\dfrac{E_a}{R_g(T+273.15)}\right)$	Okra	Doymaz (2005)
4.	$D_{eff} = 0.014\exp\left(-\dfrac{4099.8}{T_a}\right)$ where the constant before the exponential term represents diffusivity at infinitely high temperature	Hazelnuts	Özdemir & Onur Devres (1999)
5.	$D_{eff} = a \exp\left(bM^*\right)\exp\left(-\dfrac{c}{T+273.15}\right)$ where M^* represents the dimensionless moisture content	Banana	Da Silva et al. (2014)
6.	$D_{eff} = D_0 \exp\left(-\dfrac{L}{2RT} - \dfrac{K_1 K_2^{\frac{u}{u_m}}}{2T}\right)$ where u represents the local moisture concentration	Parboiled rice	Elbert et al. (2001)
7.	$D_{eff} = D_0 \exp\left(-\dfrac{E_a}{R_g T_a}\right)\exp\left(-(AT_a+B)X\right)$	Grapes	Azzouz et al. (2002)

(Continued)

TABLE 1.1 (*Continued*)
Diffusion Coefficient Relations for Selected Materials

	Effective Diffusion Coefficient	Materials	Reference
8.	$D_{eff} = D_0 \exp\left(-\dfrac{b}{273.15 + T_m} + cM\right)$ where T_m is the material temperature and M is the moisture content	Carrots	Zielinska & Markowski (2007)
9.	$D_{eff} = 8.23 \ x \ 10^{-11} \exp\left(0.0646 \ X^*\right)$ where X^* represents moisture	Squid	Teixeira & Tobinaga (1998)
10.	$D_{eff} = 0.05915 \ V^{-0.85} \exp\left(-\dfrac{4706}{V^{0.076} T}\right)$ where V describes the air speed	Rough rice	Ece & Cihan (1993)
11.	$D_{eff} = \left(C'_{10} + \varepsilon \dfrac{C'_{20}}{P_t}\right) \exp\left(-\dfrac{6.0 \exp(-20X) + E_a}{RT}\right)$ where C' are parameters characterizing the vapor and liquid contributions, P is pressure, X is moisture content and ε is porosity of matrix	Pasta	Waananen & Okos (1996)
12.	$D_{eff} = \dfrac{4.9817 \ x \ 10^{-2}}{\left(\dfrac{t}{t_h^2}\right)_{\frac{1}{2}}}$ where t is time, t_h is thickness	Pasta	Bruce Litchfield & Okos (1992)
13.	$D_{eff} = \dfrac{\pi L^2}{4t}\left(\dfrac{M_o - \bar{M}}{M_o - M_e}\right)^2$ where M_o is initial moisture, $\bar{M}$ is the average moisture content and M_e is equilibrium moisture content	Wood	Nadler et al. (1985)
14.	$D_{eff} = D_0 \exp\left(-\dfrac{E_a}{RT}\right) \exp\left((AT + B)X\right)$	Wood species	Azzouz et al. (2018)
15.	$D_{eff} = 7.18 \ x10^{-5} \exp\left(-\dfrac{31580}{RT}\right) \exp\left((-0.0025T + 1.22)X\right)$ where X is the water content	Potato	Hassini et al. (2007)
16.	$D_{eff} = D_1\left(\alpha_0 + \dfrac{1 - \alpha_0}{1 + \alpha_1^n\left(1 - H'\right)^n}\right),$ where $\alpha_1 = \frac{1 - H_{en}}{1 - H_c}$, D_1 is diffusion coefficient at $H = 1$, H is pore humidity and H_e is environmental humidity	Concrete	Bažant & Najjar (1971)
17.	$\dfrac{D_{eff}}{R^2} = A \exp\left(-\dfrac{B}{273.15 + T}\right) \exp(C \ u)$ where A, B, C are material-dependent parameters, R is radius, T is temperature of grain and u is local moisture content	Corn	Parti & Dugmanics (1990)

function of air or material temperature, following the Arrhenius-type relations, i.e., $D_{eff} = A\exp(-B/T)$, but some also take other parameters into account. Teixeira and Tobinaga (1998) investigated the behavior of the diffusion coefficient with the variation of moisture of squid muscle. In their analysis of single droplet drying using effective diffusion model, Werner et al. (2008) employed a diffusion coefficient equation developed by soil scientists (described in Bronlund 1997) with a constrictivity/tortuosity factor that accounted for the structural influence of the pore system on reducing the water vapor diffusivity in the air. Dincer and Dost (1995) developed an analytical model to determine moisture diffusivities in drying solid geometrical objects. Knowing that the governing Fickian equation is in the form of the Fourier equation of heat transfer, two different conditions for the unsteady moisture diffusion were considered, namely where Biot number takes the values 0.1<Bi<100 and Bi>100. The obtained correlation for diffusivity coefficient accounts for the characteristic length of the objects. Waananen and Okos (1996) have developed a model to correlate the effective diffusivity of moisture in extruded pasta during drying as affected by porosity, temperature and pressure.

Models of drying must consider heat and mass transfer phenomena through the boundary layer. Knowledge of the mass transfer coefficient (h_m) is required for the characterization of the boundary conditions of the heat and mass transfer equations of a liquid diffusion drying model. Dimensionless parameters are often used to correlate convective transfer data. The Sherwood number (Sh) is the dimensionless group for convective mass transfer, which represents the ratio between mass transfer by convection and mass transfer by diffusion. The mass transfer coefficients can be obtained from the definition of the Sherwood number:

$$Sh = \frac{h_m L}{D} \tag{1.6}$$

where L is a geometry-dependent characteristic length and D is the mass diffusivity. The Sherwood number can be computed using various correlations obtained experimentally or from solutions of the boundary layer equations. These correlations are functions of the Reynolds number and they differ depending upon the geometry of the objects and types of flow (laminar/turbulent, external/internal etc.).

1.2.2 VARIATIONS OF THE DIFFUSION MODEL

The complexity of the diffusion model, as mentioned above, can be adjusted as needed in the effort to describe drying as accurately as possible. Often, to simplify calculations assumptions are made when applicable, as long as they are chosen carefully so that the model does not deviate too far. The following are a few variations of assumptions and/or methods adopted in literature. The major assumption widely employed in determining diffusivity is that drying is mass transfer limited and the temperature remains constant throughout drying. Many have pointed out that achieving this isothermal condition is essential since the presence of a temperature gradient may result in discrepancies between predictions and experimental data. Srikiatden and Roberts (2006) compared effective diffusivities measured traditionally using convective hot air (non-isothermal) to those measured under isothermal conditions. The isothermal drying apparatus used combined microwave energy with convective hot air and provided the correct temperature-dependent diffusion coefficient. Using the same approach, two years later these authors predicted the drying behavior in hygroscopic materials during convective drying by using the isothermally determined effective diffusivity in a simultaneous heat and mass transfer (SMHT) model based on Fick's law. Chen, Zheng, and Zhu (2012 analyzed the drying characteristics of poplar sawdust using a thermogravimetric analyzer (TGA) to prevent temperature gradients. The isothermal condition is very difficult to maintain practically, thus alternative approaches with non-isothermal conditions are also of great interest. Li and Kobayashi (2005) as well as Chen, Zhang, and Zhu (2012) numerically determined moisture diffusivity with the experiments conducted in a TGA and compared the characteristics of isothermal and non-isothermal procedures.

Most studies rely on the assumption of an isotropic solid, resulting in a unique water diffusivity for all directions. However, anisotropic behavior has been confirmed for various materials and effective diffusivities identification methods have been developed by multiple authors. Adamski and Pakowski (2013) estimated radial, axial and angular water diffusivities during pine wood drying with superheated steam by numerically solved 3D heat and mass transfer equations. Pacheco-Aguirre et al. (2014) have proposed a method to estimate moisture diffusivities in radial, axial and angular directions during drying of anisotropic cylindrical solids, using carrots as samples. The developed method was based on the analytical solution for the non-steady mass transfer equation in products shaped as longitudinal sections of finite anisotropic cylinders. Souraki and Mowla (2008) adjusted experimental drying curves to Fick's law for infinite and finite cylinders with and without consideration of shrinkage to measure axial and radial moisture diffusivities of cylindrical green beans in a fluidized bed dryer.

Moreover, many of the diffusion models do not take full account of the influence of microscale effects, like capillary effect, Kelvin effect and microscopic structure. In the past decades, most of the authors have included these effects, which are usually lumped into the effective diffusion coefficient. In addition, heat transfer phenomena can also be taken into account into the model. However, this increases the computational complexity and cost.

In 1985, Plumb et al. recognized two major impediments to the development of predictive models for transport in wood drying: lack of detailed experimental data quantifying drying rates and moisture and temperature distribution, and little attention towards the necessary transport properties. To overcome these deficiencies, they suggested a model for heat and mass transport during wood drying in which the transport properties are developed from knowledge of wood structure using a mechanistic model. The model describes transfer of liquid via both capillary and diffusive transport. Six years later, Dolinskiy et al. (1991) studied the conjugate heat and mass exchange problem in the initial period of drying of a continuously moving material. The problem considered the influence of processes taking place within the material on the heat and mass transfer rate, which otherwise cannot be considered by traditional calculation techniques. It was found that the solution of the conjugate problem in convective drying yields decreased drying rates compared to those obtained by traditional methods. While the results were obtained under the assumption of laminar flow regime, it can be demonstrated that they are also qualitatively valid for turbulent flow regime and for real drying. The established decrease in the rate will take place in any process under the conditions of falling concentration head. Recent research has been able to include more phenomena into the models with the help of technology. Niamnuy et al. (2008) formulated a model to simulate the transient, three-dimensional heat and moisture transfer during drying of irregularly shaped biomaterial, namely shrimp, in a jet-spouted bed dryer. The model considered the mechanical deformation and was also used to predict the principal stress distributions within shrimp during drying. Using the same approach, Lemus-Mondaca et al. (2013) carried out a numerical study for the drying process of papaya slices to monitor the unsteady temperature and moisture distributions inside the sample. Recently, Yuan et al. (2018) proposed the molecular dynamics model of moisture diffusion and used the molecular dynamics (MD) simulation method to simulate and analyze the drying process of moisture diffusion in nanopores of porous media, with consideration of the Kelvin effect.

1.3 POROUS MEDIA MODEL

Like the diffusion model, the porous media model is a continuum-based method, where the porous medium is treated as a continuous material that allows the measurement of quantities of interest over volumes or areas that cross many pores. This approach describes the flow through porous media through the application of volume averaging to obtain macroscopic quantities equations. In the spatial averaging approach, the variable is defined as an appropriate mean over a sufficiently large *representative elementary volume* (REV). The length scale of the REV is larger than the pore scale, but much smaller than the flow domain (Nield and Bejan 2017). Figure 1.2 shows the volume V constituted of three different phases: a matrix solid, a liquid phase and a gaseous phase. However, for single-phase flow through solid matrices, the volume is assumed to be occupied by solid and fluid only. The average value of any quantity ψ at every point in space (both in the solid phase and fluid phase) is defined by

$$\psi_\beta = \frac{1}{V} \int_{V_\beta} \psi_\beta \, dV \tag{1.7}$$

where V_β represents the volume of the β-phase contained within the averaging volume.

The actual fluid velocity varies throughout the pore space due to the connectivity and geometric complexity of that space. This velocity can be characterized by its mean or average value, called the average fluid velocity. From this point onwards, the superficial (Darcian) velocity $\mathbf{u}$ will be used, which is defined as

$$\mathbf{u} = \varepsilon \mathbf{u}_f \tag{1.8}$$

where ε represents porosity and $\mathbf{u}_f$ is the fluid velocity.

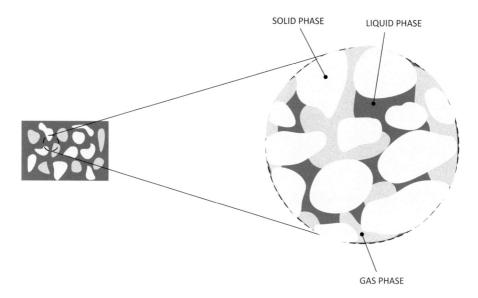

FIGURE 1.2   Schematic of a representative elementary volume (Adapted from Yazbek et al. 2006).

1.3.1   Single-Phase Flow through Porous Media

In single-phase flow through porous media, the liquid transport is characterized by the conservation of mass, momentum and energy equations formulated as follows (Bird 2007):

- Conservation of mass

$$\frac{\partial}{\partial t}(\rho) + \nabla \cdot (\rho \mathbf{u}) = -R_{evap} \tag{1.9}$$

where R_{evap} is the rate of evaporation of liquid water to vapor per unit volume.

- Conservation of momentum

$$\frac{1}{\varepsilon}\frac{\partial}{\partial t}(\rho \mathbf{u}) + \frac{1}{\varepsilon^2}\left[\nabla \cdot (\rho \mathbf{uu})\right] = \frac{1}{\varepsilon}\nabla \cdot \left(\mu\left(\nabla \mathbf{u} + \nabla \mathbf{u}^T\right)\right) - \nabla P + S \tag{1.10}$$

where S is the source term making up the total resistance to the flow. The source term is composed of three parts: a viscous loss term (Darcy source), an inertial (Ergun) loss term, and capillary force term

$$S_D = -\frac{\mu}{K}\mathbf{u} \tag{1.11}$$

$$S_E = -\frac{C_E}{K^{\frac{1}{2}}} \rho |\mathbf{u}| \mathbf{u} \tag{1.12}$$

$$F_c = \sigma_{lg} \frac{\rho \kappa_l \nabla \alpha_l}{0.5 \left(\rho_l + \rho_g \right)} \tag{1.13}$$

$$S = S_D + S_E + F_c \tag{1.14}$$

where K represents the permeability of the solid matrix, C_E stands for the Ergun coefficient, σ is the surface tension coefficient, α_l is the volume fraction of the liquid phase and κ is the surface curvature, defined in terms of the divergence unit normal $\hat{n}$, as follows (Brackbill et al. 1992):

$$\kappa = \nabla \cdot \hat{n} \tag{1.15}$$

$$\hat{n} = \frac{\nabla \alpha_l}{|\nabla \alpha_l|} \tag{1.16}$$

• Conservation of species i

$$\frac{\partial}{\partial t} \left(c_{liq} \right) + \left[\nabla \cdot \left(\mathbf{u} c_{liq} \right) \right] = D_{liq} \nabla^2 c_{liq} + S_{liq} \tag{1.17}$$

where c_{liq} is liquid concentration, D_{liq} is liquid (water) diffusivity and S_{liq} is the liquid source which, in the case of drying, is essentially the rate of evaporation as in Eq. 1.9.

• Conservation of energy

The average temperature over a representative volume element, RVE, for solid phase and fluid phase in a porous material may be significantly different. The local rate of change of temperature for the two phases will not be equal when the temperature at the bounding surface changes with time and both phases have significantly different heat capacities C_p and thermal conductivities (Kaviany 1999). The temperature variations of fluid and solid phases induce additional pore and thermal stresses in the porous medium (Gandomkar and Gray 2018). This situation, known as local thermal non-equilibrium (LTNE), leads to application of the LTNE approach, which generates two energy equations for each phase coupled at the solid-fluid interface A_{sf}. The energy equation as it applies to the solid and fluid in the pores, written for each phase, is

$$(1-\varepsilon)\frac{\partial}{\partial t}\left(\rho_s h_s\right) = \nabla \cdot \left((1-\varepsilon)k_s \nabla T_s\right) + h_{sf} A_{sf} \left(T_s - T_f\right) \tag{1.18}$$

$$\varepsilon \frac{\partial}{\partial t}\left(\rho_f h_f\right) + \nabla \cdot \left(\rho_f h_f \mathbf{u}\right) = \nabla \cdot \left(\varepsilon k_f \nabla T_f\right) + h_{sf} A_{sf} \left(T_f - T_s\right) + S_l \tag{1.19}$$

where enthalpy h for each phase can be approximated by $h = C_p T$, h_{sf} is the interstitial convection heat transfer coefficient and S_l is source of latent heat evaporation if there is any mass sink.

As seen above, the single energy equation needs to be replaced with two energy equations in the absence of local thermal equilibrium (LTE). The LTE approach, which assumes that local-averaged temperatures of solid, T_s, and fluid, T_f, are equal, is prevalent in modeling transport phenomena in porous media (Mohamad 2003; Jang and Chen 1992; Chikh et al. 1995). The local thermal equilibrium approximation is valid when the time scale t satisfies (Kaviany 1999):

$$\frac{\varepsilon(\rho C_p)_f \, \ell^2}{t}\left(\frac{1}{k_f}+\frac{1}{k_s}\right) \ll 1 \tag{1.20}$$

$$\frac{(1-\varepsilon)(\rho C_p)_s \, \ell^2}{t}\left(\frac{1}{k_f}+\frac{1}{k_s}\right) \ll 1 \tag{1.21}$$

and length scale satisfy (Kaviany 1999):

$$\frac{\varepsilon k_f \ell}{A_0 L^2}\left(\frac{1}{k_f}+\frac{1}{k_s}\right) \ll 1 \tag{1.22}$$

$$\frac{(1-\varepsilon)k_s \ell}{A_0 L^2}\left(\frac{1}{k_f}+\frac{1}{k_s}\right) \ll 1 \tag{1.23}$$

where A_0 is the specific surface area.

Additionally, Minkowycz et al. (1999) have confirmed the use of the Sparrow number S_p to ascertain the LTE condition and concluded that a large S_p ($S_p > 500$) is indicative of the presence of a local thermal equilibrium condition.

$$S_p = Nu\left(\frac{k_f}{k_e}\right)\left(\frac{L}{r_h}\right)^2 \tag{1.24}$$

A single LTE conservation equation of energy can therefore be built upon the above definitions by combining Eqs. 1.18 and 1.19 (Nield and Bejan 2017):

$$\frac{\partial}{\partial t}(\rho h)_m + \nabla\cdot\left((\rho h)_f \mathbf{u}\right) = \nabla\cdot(k_m \nabla T) + S_H \tag{1.25}$$

where

$$(\rho h)_m = (1-\varepsilon)(\rho h)_s + \varepsilon(\rho h)_f \tag{1.26}$$

$$k_m = (1 - \varepsilon)k_s + \varepsilon k_f \qquad (1.27)$$

are the overall (mixture) heat capacity per unit volume and overall thermal conductivity, respectively. The first term on the left-hand side of Eq. 1.25 is a transient term which represents the rate of accumulation of energy, while the second term describes the heat transfer due to convection. On the right-hand side of the equation is a conduction term. Typically, there is no external heat source in drying, but one could include the heat source term S_H when there is latent heat of evaporation (mass sink) or when employing microwave drying, which involves a volumetric heat source due to electric and magnetic potential.

In describing flow through porous medium, determining parameters like permeability is as important as solving the conservation equations. These properties are commonly defined by constitutive equations. Accurate measurement of permeability is critical for fluid flow modeling in porous materials. Sharma and Siginer (2010) have provided a review of various experimental methods to measure permeability. Bryant and Blunt (1992) proposed a predictive calculation of two-phase relative permeabilities for unconsolidated well-sorted sandstone modeled as monodisperse spheres with random close packing. The porous media was formed from a dense random packing of equal spheres, allowing the microstructure of the medium to be completely determined once the spatial coordinates of every sphere was measured. Such packing gives the flexibility to transform the model into that of a consolidated sandstone by allowing the spheres to swell. The pore structure was represented by a network model, from which the relative permeabilities could be predicted; the results compared well with experimental measurements on monomineralic, well-sorted materials. This method enables calculation of relative permeabilities without supplementary measurements on the sample. Bosl et al. (1998) numerically calculated the absolute (single-phase) permeability of simulated granular rocks by explicitly modeling Stokes flow using a lattice Boltzmann method. Although many constitutive relations have been proposed, the most widely used relation for permeability K is the Kozeny equation (Nield and Bejan 2017):

$$K = \frac{d_p^2 \varepsilon^3}{\beta(1 - \varepsilon)^2} \qquad (1.28)$$

where d_p is the particle diameter and β is an empirically determined shape factor. Beavers et al. (1973) have performed experiments to determine the influence of the presence of container walls on the value of the inertial loss coefficient (Eq. 1.12) and found that the value is substantially influenced by the presence of bounding walls, following the expression:

$$C_E = 0.55\left(1 - 5.5\frac{d}{D_e}\right) \qquad (1.29)$$

where d is the spheres diameter and D_e is the equivalent diameter of the bed, defined in terms of the width w and height h of the bed as:

$$D_e = \frac{2wh}{w+h} \qquad (1.30)$$

1.3.2 Multiphase Flow through Porous Media

In order to better capture drying of porous materials, one needs to consider the simultaneous heat and mass transfer in all phases. One key distinction of multiphase flow in porous media when compared to the single-phase flows is the wetting of the surface of the matrix by one of the fluid phases. The hydrodynamics of multiphase flow in porous media is in part controlled by the dynamics of the liquid-gas-solid contact line (Rohsenow et al. 1998). The presence of a curvature at the liquid-gas interface results in difference between the local gaseous and liquid phase pressures (capillary pressure), which depends on how much the wetting phase occupies the volume and will affect the heat and mass transfer. The following section will focus on a two-phase system, although the equations presented can be easily extended to more than two-phase flow.

1.3.2.1 Single-Fluid Model

The first approximation for the analysis of two-phase flows is the single-fluid approach. A typical model used in this approach is the mixture model, which assumes that individual fluid phases behave as a flowing mixture described in terms of the mixture properties. The mixture model is a sufficiently accurate approximation to model multiphase flows where the phases move at different velocities, with only a moderate increase in the computational effort compared to the single-phase approach. The model solves the mixture conservation equations, the secondary-phase volume fraction equation and establishes relative velocities for the dispersed phases (El-Batsh et al. 2012). For a mixture of n phases and volume fraction of phase k, α_k, the continuity equation is (Nield and Bejan 2017; Kaviany 1999):

$$\frac{\partial}{\partial t}(\rho_m) + \nabla \cdot (\rho_m \mathbf{u}_m) = 0 \qquad (1.31)$$

where $\mathbf{u}_m$ and ρ_m are the mass-averaged velocity, determined based on the volume-averaging method, and the mixture density, respectively. When applied to drying, the two phases present are liquid (l) as primary phase and gaseous (g) as secondary phase, and therefore the mixture velocity and density can be defined as:

$$\mathbf{u}_m = \frac{1}{\rho_m} \sum_{k=1}^{n} \alpha_k \rho_k \mathbf{u}_k = \frac{1}{\rho_m} \left(\alpha_l \rho_l \mathbf{u}_l + \alpha_g \rho_g \mathbf{u}_g \right) \qquad (1.32)$$

$$\rho_m = \sum_{k=1}^{n} \alpha_k \rho_k = \alpha_l \rho_l + \alpha_g \rho_g \qquad (1.33)$$

The momentum equation for the mixture is obtained by summing the individual momentum equations for all phases (Nield and Bejan 2017; Kaviany 1999):

$$\frac{1}{\varepsilon}\frac{\partial}{\partial t}\left(\rho_m\mathbf{u}_m\right)+\frac{1}{\varepsilon^2}\left[\nabla\cdot\left(\rho_m\mathbf{u}_m\mathbf{u}_m\right)\right]$$

$$=\frac{1}{\varepsilon}\nabla\cdot\left(\mu_m\left(\nabla\mathbf{u}+\nabla\mathbf{u}^T\right)\right)-\nabla P+\rho_m\vec{g}+S+\frac{1}{\varepsilon^2}\nabla \qquad (1.34)$$

$$\cdot\left(\alpha_l\rho_l\mathbf{u}_{d,l}\mathbf{u}_{d,l}+\alpha_g\rho_g\mathbf{u}_{d,g}\mathbf{u}_{d,g}\right)$$

where S is the source term as defined in Eqs. 1.11–1.14, the viscosity of the mixture μ_m is given by:

$$\mu_m=\sum_{k=1}^{n}\alpha_k\rho_k=\alpha_l\rho_l+\alpha_g\rho_g \qquad (1.35)$$

and the drift velocity for gaseous phase g, $\mathbf{u}_{d,g}$, in the interface momentum transfer term (last term on the right-hand side) is the difference between the velocity of the dispersed phase and the mixture velocity.

$$\mathbf{u}_{d,g}=\mathbf{u}_g-\mathbf{u}_m \qquad (1.36)$$

The energy equation in the mixture model takes the following form:

$$\frac{\partial}{\partial t}\left(\rho_m E_m\right)+\nabla\cdot\left(\rho_m\mathbf{u}_m E_m\right)=\nabla\cdot\left(k_{eff}\nabla T\right)+S \qquad (1.37)$$

The first term on the right-hand side of Eq. 1.37 represents energy transfer due to conduction and it is dependent on the effective conductivity k_{eff}, which is a function of the thermal conductivity of the fluid and of the porous solid (k_s).

$$k_{eff}=\varepsilon\left(\alpha_l k_l+\alpha_g k_g\right)+(1-\varepsilon)K_s \qquad (1.38)$$

The relative velocity $\mathbf{u}_{gl}$, also called the slip velocity, is defined as the velocity of the gaseous phase g relative to the velocity of the liquid phase l.

$$\mathbf{u}_{gl}=\mathbf{u}_g-\mathbf{u}_l \qquad (1.39)$$

The relative velocity can be calculated by (Manninen et al. 1996):

$$\mathbf{u}_{gl}=\frac{d_g^2\left(\rho_g-\rho_m\right)}{18\mu_l F_D}\vec{a} \qquad (1.40)$$

where d_g is the gas bubble diameter and $\vec{a}$ is the gas bubble or water vapor acceleration, given by:

$$\vec{a} = \vec{g} - \left(\mathbf{u}_m \cdot \nabla\right)\boldsymbol{u}_m - \frac{\partial}{\partial t}\left(\mathbf{u}_m\right) \tag{1.41}$$

Many correlations for drag force F_D have been proposed. The most commonly used one is the standard drag force correlations proposed by Schiller-Naumann (Schiller & Naumann 1935):

$$F_D = \begin{cases} 1 + 0.15Re^{0.687} & Re \le 1000 \\ 0.0183Re & Re > 1000 \end{cases} \tag{1.42}$$

The volume fraction equation for gas phase, derived from the continuity equation, takes the following form:

$$\frac{\partial}{\partial t}\left(\alpha_g \rho_g\right) + \nabla \cdot \left(\alpha_g \rho_g \mathbf{u}_m\right) = -\nabla \cdot \left(\alpha_g \rho_g \mathbf{u}_{d,g}\right) + \sum_{p=1}^{n}\left(\dot{m}_{lg} - \dot{m}_{gl}\right) \tag{1.43}$$

1.3.2.2 Multi-Fluid Model

The governing equations for single-fluid flow for both phases are Eqs. 1.31, 1.34 and 1.37, which were formulated based on the assumption of a binary mixture. Modeling of multiphase flow through a porous medium becomes more intricate when two or more fluids coexist. In the multi-fluid model, a set of conservation equations for each phase is solved independently and the interaction between phases is coupled in the interphase mass, momentum and energy transfer, which further complicates attempts to describe the fluid movement mathematically. The multiphase, multi-fluid model has been commonly used in an effort to model drying of porous media at the macroscopic level. The model allows better comprehension of the transport mechanism in drying of porous media because it considers the transport mechanism of liquid water, vapor and air separately. Description of fluid interaction such as capillary pressure and relative permeability is essential and must be introduced into the equations to capture a better model for multiphase flow.

When two fluids coexist in a porous medium, one fluid will have preferential wettability for the solid phase and will occupy smaller voids as compared to the nonwetting fluid (generally gaseous phase), as shown in Figure 1.3 (Kamyabi 2014). Pressure difference across the interface, called capillary pressure, will then occur, owing to the presence of curvature at the interface; this difference depends on the fraction occupied by the wetting phase (saturation) s.

In addition, the flow of one fluid is susceptible to the presence of other fluids, which means that permeability of one fluid is affected by properties of the other fluids. Many researchers relate relative permeability to capillary pressure. Relative permeability of

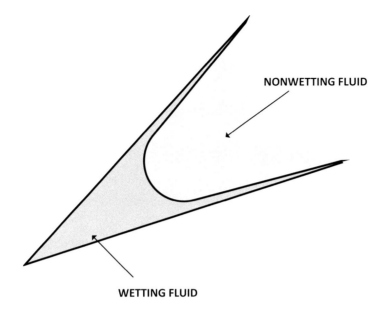

NONWETTING FLUID

WETTING FLUID

FIGURE 1.3 Schematic of gas – liquid interphase for wetting and nonwetting fluid.

the wetting phase (liquid) K_{rl} and that of the nonwetting phase (gaseous) K_{rg} can be calculated using the Purcell Relative Permeability Model as follows (Purcell 1949):

$$K_{rl} = \left(s_l^* \right)^{\frac{2+\lambda}{\lambda}} \tag{1.44}$$

$$K_{rg} = \left[1 - \left(s_l^* \right)^{\frac{2+\lambda}{\lambda}} \right] \tag{1.45}$$

where s_l^* is the normalized liquid phase saturation and λ is the pore distribution index. Similarly, for drainage cases with liquid (oil) as the wetting phases and gas as nonwetting phase, Corey (1954) proposed the Corey Relative Permeability Model.

$$K_{rl} = \left(s_l^* \right)^4 \tag{1.46}$$

$$K_{rg} = \left(1 - s_l^* \right)^2 \left[1 - \left(s_l^* \right)^2 \right] \tag{1.47}$$

Li and Horne (2006) have compared various relative permeability calculation methods to verify which capillary pressure model yields the best fit to experimental data. Several mathematical models have also been proposed to estimate relative permeabilities from data of other parameters. Li (2008) developed a method to infer relative permeabilities from resistivity index data. The method used empirically determined relative permeabilities of the wetting phase to obtain the nonwetting phase permeabilities. The proposed equation to calculate the relative permeability of the wetting phase according to the analogy between fluid flow and electrical flow is:

$$K_{rl} = \frac{R_o}{R_t} = \frac{1}{I} \tag{1.48}$$

where R_t is the real resistivity, R_o is the resistivity of the same material at 100% water saturation and I is the resistivity index. A numerical simulation to predict the behavior of relative permeability in a porous body with immobile fluid inclusions (obstacles) using two-scale modeling was carried out by Markicevic and Djilali (2006). The results showed that relative permeability varies according to a power law of saturation, with quasilinear behavior for low permeability and nonlinear behavior for higher permeability.

The mathematical model consists of conservation equations for all the phases present. The general continuity equation for liquid and gas phases, based on the Eulerian approach, is as follows (Nield and Bejan 2017).

$$\frac{\partial}{\partial t}(\alpha_l \rho_l) + \nabla \cdot (\alpha_l \rho_l \mathbf{u}_l) = \dot{m}_l + S_l \tag{1.49}$$

$$\frac{\partial}{\partial t}(\alpha_g \rho_g) + \nabla \cdot (\alpha_g \rho_g \mathbf{u}_g) = \dot{m}_g + S_g \tag{1.50}$$

The first term on the right-hand side of both equations represents the interphase mass transfer from liquid to gas or vice versa and the second term represents an additional volumetric source of mass term, which is usually set to zero when no chemical reaction or radiation is considered. In drying, specifically, the interphase mass transfer could be a consequence of evaporation and/or condensation.

The momentum balance equations for liquid and gas phases are described by the following equations (Rohsenow et al. 1998):

$$\frac{\partial}{\partial t}(\alpha_l \rho_l \mathbf{u}_l) + \nabla \cdot (\alpha_l \rho_l \mathbf{u}_l \mathbf{u}_l) = -\alpha_l \nabla P + \nabla \cdot \overline{\overline{\tau}}_l + \alpha_l \rho_l \vec{g} - \frac{\mu_l}{K_l} \mathbf{u}_l + \vec{R} + \vec{F} \tag{1.51}$$

$$\frac{\partial}{\partial t}(\alpha_g \rho_g \mathbf{u}_g) + \nabla \cdot (\alpha_g \rho_g \mathbf{u}_g \mathbf{u}_g) = -\alpha_g \nabla P + \nabla \cdot \overline{\overline{\tau}}_g + \alpha_g \rho_g \vec{g} - \frac{\mu_g}{K_g} \mathbf{u}_g + \vec{R} + \vec{F} \tag{1.52}$$

where $\overline{\overline{\tau}}_l, \overline{\overline{\tau}}_g$ describes the total stress-strain tensor of liquid and gas phase respectively (defined in Eqs. 1.55–1.56), P is the static pressure of the mixture, K_l and K_g is the liquid phase permeability and gaseous phase permeability, $\vec{R}$ is the momentum exchanged between phases and $\vec{F}$ is the net additional forces affecting the phases. The phase permeability used in the multiphase flow momentum equation is the corrected permeability, defined using the relative permeabilities in the following way (Muskat and Meres 1936)

$$K_l = K K_{rl} \tag{1.53}$$

$$K_g = KK_{rg} \tag{1.54}$$

where K is the absolute permeability of the porous material as used in the single-phase flows and K_r is the relative permeability of the corresponding phase, which can be obtained from Eqs. 1.44–1.47.

The stress-strain tensor is defined by (Hirsch 2007):

$$\bar{\bar{\tau}}_l = \alpha_l \rho_l \left(\nabla \mathbf{u}_l + \nabla \mathbf{u}_l^T \right) + \alpha_l \left(\lambda_l - \frac{2}{3} \mu_l \right) \nabla \cdot \mathbf{u}_l \bar{\bar{I}} \tag{1.55}$$

$$\bar{\bar{\tau}}_g = \alpha_g \rho_g \left(\nabla \mathbf{u}_g + \nabla \mathbf{u}_g^T \right) + \alpha_g \left(\lambda_g - \frac{2}{3} \mu_g \right) \nabla \cdot \mathbf{u}_g \bar{\bar{I}} \tag{1.56}$$

where μ and λ are the shear and bulk viscosity of the phase and $\bar{\bar{I}}$ is the unit tensor (identity matrix).

Momentum exchange between the liquid and gaseous phases is given by

$$\vec{R} = \vec{R}_{lg} + \dot{m}_{lg} \mathbf{u}_{lg} - \dot{m}_{gl} \mathbf{u}_{gl} \tag{1.57}$$

where $\vec{R}_{lg}$ is the interphase force between liquid and gas phase and $\mathbf{u}_{lg}$ represents the interphase velocity, which depends on the direction of the mass transfer:

$$\text{if } \dot{m}_{lg} > 0, \ \mathbf{u}_{lg} = \mathbf{u}_l, \text{ and if } \dot{m}_{lg} < 0, \text{ then } \mathbf{u}_{lg} = \mathbf{u}_g.$$

It is important to mention that for modeling multiphase flows, these exchange terms cause the majority of uncertainties, because they include parameters that are formulated relying on experiments and/or mathematical assumptions. The interphase force $\vec{R}_{lg}$ can be related to the interphase momentum exchange coefficient C_{lg} and both phase velocities by:

$$\vec{R}_{lg} = C_{lg} \left(\mathbf{u}_l - \mathbf{u}_g \right) \tag{1.58}$$

Since momentum transfer occurs between phases, one phase will gain while the other one loses, therefore the dependences $\vec{R}_{lg} = -\vec{R}_{gl}$, $C_{lg} = C_{gl}$, and $\vec{R}_{gg} = 0$ must be satisfied. Many different models describing the coefficient of momentum exchange C_{lg} exist. For a fluid-fluid system like in drying where there are unequal amounts of fluids, liquid water, being the predominant fluid, is taken as the primary phase fluid and the exchange coefficient takes the form (Ubbink 1997; Muzaferija et al. 1999):

$$C_{lg} = \frac{3\mu_g f}{d_l} A_i \tag{1.59}$$

where d_l is the characteristic length of the pore, A_i is the interfacial area between liquid and gas and f, the drag function, is dependent on the Reynolds number (Re)

TABLE 1.2

Drag Coefficient Models

Model

Schiller-Naumann (1935)

$$f = \frac{C_D Re}{24}$$

$$C_D = \begin{cases} 24\left(1 + 0.15 Re^{0.687}\right)/Re & Re \leq 1000 \\ 0.44 & Re > 1000 \end{cases}$$

Morsi-Alexander (1972)

$$f = \frac{C_D Re}{24}$$

$$C_D = a_1 + \frac{a_2}{Re} + \frac{a_3}{Re}$$

	a_1	a_2	a_3
$0 < Re < 0.1$	0	24	0
$0.1 < Re < 1$	3.690	22.73	0.0903
$1 < Re < 10$	1.222	29.1667	−3.8889
$10 < Re < 100$	0.6167	46.50	−116.67
$100 < Re < 1000$	0.3644	98.33	−2778
$1000 < Re < 5000$	0.357	148.62	−47,500
$5000 < Re < 10,000$	0.46	−490.546	5,78,700
$Re > 10,000$	0.5191	−490.546	54,16,700

Symmetric
(Bowen, 1976)

$$C_{lg} = \frac{3\left(\alpha_l \mu_l + \alpha_g \mu_g\right) f}{d_{lg}} A_i$$

$$d_{lg} = \frac{1}{2}\left(d_l + d_g\right)$$

$$f = \frac{C_D Re}{24}$$

$$C_D = \begin{cases} 24\left(1 + 0.15 Re^{0.687}\right)/Re & Re \leq 1000 \\ 0.44 & Re > 1000 \end{cases}$$

$$Re = \frac{\left(\alpha_l \rho_l + \alpha_g \rho_g\right)\left|\mathbf{u}_l - \mathbf{u}_g\right| d_{lg}}{\alpha_l \mu_l + \alpha_g \mu_g}$$

and is defined differently for different models. Table 1.2 lists a few examples of drag models that can be used for two-fluid CFD simulations.

The Schiller-Naumann model is one of the most widely used drag models. This model assumes that the drop is immersed in an infinite pool of the continuous phase and that effects induced by the presence of nearby drops are negligible (Sen et al. 2016). The Morsi and Alexander model has a different functional form from the Schiller-Naumann model, adjusting the function definition frequently over a large range of Reynolds numbers. In both the Schiller-Naumann and the Morsi-Alexander models, the drag force depends on the relative Reynolds number (Re). The relative Reynolds number is calculated from

$$Re = \frac{\rho_l \left| \mathbf{u}_l - \mathbf{u}_g \right| d_l}{\mu_g} \tag{1.60}$$

The symmetric model has the same drag coefficient relationship as Schiller-Naumann but the Reynolds number is defined with respect to the mixture:

$$Re = \frac{\rho_{lg} \left| \mathbf{u}_l - \mathbf{u}_g \right| d_{lg}}{\mu_{lg}} \tag{1.61}$$

The third term on the right-hand side of Eqs. 1.51–1.52 describes the body force acting on the fluid and the last term in the momentum balance describes the total additional source forces influencing the fluid. The following are a few examples of the interphase forces that could be significant in drying.

1. Lift Force

   Lift force, the force on a body moving relative to a straining fluid, is caused by the shearing effect of the fluid onto the particle. In most cases, lift force is insignificant compared to drag force and thus negligible, except when the phases separate quickly. In drying processes, this depends upon the heat flux intensity. If the heat flow rate is high enough to create a gas bubble, lift force can have a notable effect. On the other hand, when the heat flux is low, i.e., drying is slow, nonhomogeneous drying and hence, gas bubble formation, can be avoided and lift force effect need not be considered. The lift force can be described as follows (Drew & Lahey 1990):

   $$\vec{F}_{lift,g} = -C_{lift}\rho_g\alpha_l \left(\mathbf{u}_g - \mathbf{u}_l \right) \times \left(\nabla \times \mathbf{u}_g \right) \tag{1.62}$$

   and the equity $\vec{F}_{lift,g} = -\vec{F}_{lift,l}$ must be satisfied. Auton (1987) and Drew and Lahey (1987,1990) have calculated the lift force for the case of a rigid sphere and their analyses yield $C_{lift} = 0.5$. However, this hypotheses is considered inaccurate because the calculation did not take into account wake effects and considered the coefficient as a constant independent of local average velocity and average vorticity. According to Moraga et al. (1999), in practice, the lateral force experienced by a particle will depend on the instantaneous velocity field around it and they defined the lift coefficient in terms of both particle Reynolds number Re_p and vorticity Reynolds number Re_v as follows (Moraga et al. 1999):

$$C_{lift} = \begin{cases} 0.0767 & \varphi \leq 6000 \\ -\left(0.12 - 0.2e^{-\frac{\varphi}{360000}} \right)e^{-\frac{\varphi}{3\times10^7}} & 6000 < \varphi < 5\times10^7 \\ -0.6353 & \varphi \geq 5\times10^7 \end{cases} \tag{1.63}$$

where

$$\varphi = Re_p Re_v \qquad (1.64)$$

$$Re_p = \frac{\rho_g \left| \mathbf{u}_g - \mathbf{u}_l \right| d_l}{\mu_g} \qquad (1.65)$$

$$Re_v = \frac{\rho_g \left| \nabla \times \mathbf{u}_g \right| d_l^2}{\mu_g} \qquad (1.66)$$

This lift coefficient combines opposing actions of two lift forces: "classical" aerodynamics lift force resulting from interaction between bubble and liquid shear, and lateral force resulting from interaction between bubble and vortices shed by bubble wake.

2. Virtual Mass Force

The effect of virtual mass plays an important role in cases where the density of the dissipated phase (i.e., water vapor) is much lower than density of the medium (i.e., liquid water) (Sobieski 2008). The so-called virtual mass force arises when a rigid body accelerates through a quiescent fluid, causing the environment to counteract and decelerate the accelerating gas bubble or water vapor. This force is defined as (Drew and Lahey 1987):

$$\vec{F}_{vm,g} = 0.5\rho_g \alpha_l \left(\frac{d_g \mathbf{u}_g}{dt} - \frac{d_l \mathbf{u}_l}{dt} \right) \qquad (1.67)$$

and is added to the right-hand side of the momentum equation for both phases: their values are identical, but they have opposite directions ($\vec{F}_{vm,g} = -\vec{F}_{vm,l}$). The derivatives d_g / dt describe the phase material time derivative and take the form:

$$\frac{d_g(\phi)}{dt} = \frac{\partial}{\partial t}(\phi) + \left(\mathbf{u}_g \cdot \nabla \right)\phi \qquad (1.68)$$

3. Capillary Force

During drying of porous media, moisture movement is largely affected by surface tension forces or by capillary action. The transport owing to capillary forces depends upon the surface curvature, as shown in Eq. 1.13.

As seen above, modeling multiphase flow through porous media leads to a very complicated set of equations and closure relations. In some cases it is possible to use a more straightforward approach, for example by deriving the momentum conservation equation from Darcy's equation (Kumar et al. 2014, 2016; Turner and Perré 2004; Feng et al. 2001), which will result in coupled mass and momentum equation, rather than solving the momentum balance independently. The mass-momentum equation for the liquid water is expressed by (Kumar et al. 2016)

$$\frac{\partial}{\partial t}\left(\varepsilon S_l \rho_l\right)+\nabla\cdot\left(\vec{n}_l\right)=-R_{evap} \tag{1.69}$$

where S_l is the liquid water saturation defined as the fraction of pore volume occupied by the liquid phase ($S_l = V_l / \varepsilon V$), $\vec{n}_l$ is the total liquid water flux due to the gradient of liquid pressure, as given by Darcy's Law:

$$\vec{n}_l = -\rho_l \frac{K_l K_{rl}}{\mu_l}\nabla P - D_c \nabla c_l \tag{1.70}$$

where P is the total gas pressure, K_l is the intrinsic permeability of liquid water, K_{rl} is relative permeability of liquid water, μ_l is the viscosity of water, D_c is capillary diffusivity and c_l is the liquid concentration. The first term in Eq. 1.70 represents the flow due to gradients in gas pressure, while the second term describes the flow due to capillary pressure. The capillary diffusivity is expressed in terms of permeabilities K and concentration-dependent capillary pressure P_c:

$$D_c = \rho_l \frac{K_l K_{rl}}{\mu_l}\frac{\partial P_c}{\partial c_l} \tag{1.71}$$

where (Chemki et al. 2009):

$$P_c = \sqrt{\frac{\varepsilon}{K_l}}\times\left(0.1212 - 0.000167T\right)\times J\left(S_l\right) \tag{1.72}$$

$$J\left(S_l\right)= 0.364\left(1-\exp\left(40\ S_l - 40\right)\right)+0.221\left(1-S_l\right)+\frac{0.005}{S_l} \tag{1.73}$$

where ε is the porosity, K_l is the intrinsic permeability of liquid, T is temperature and S_l is the liquid saturation.

The conservation of water vapor is given by (Kumar et al. 2016):

$$\frac{\partial}{\partial t}\left(\varepsilon S_g \rho_g \omega_v\right)+\nabla\cdot\left(\vec{n}_v\right)= R_{evap} \tag{1.74}$$

where S_g is the gas saturation ($S_g = 1 - S_l$), ρ_g is density of gas, ω_v is the mass fraction of vapor and $\vec{n}_v$ is the vapor mass flux, which for a binary mixture, can be written as:

$$\vec{n}_v = -\rho_g \omega_v \frac{K_g K_{rg}}{\mu_g}\nabla P - \varepsilon S_g \rho_g D_{eff,g}\nabla\omega_v \tag{1.75}$$

where K_g is intrinsic permeability of gas, K_{rg} is the relative permeability of gas, μ_g is the viscosity of gas and $D_{eff.g}$ is the binary diffusivity of vapor and air.

The gas phase is a mixture of vapor and air and its pressure P can be calculated by solving the total mass balance for the gas phase, namely,

$$\frac{\partial}{\partial t}\left(\varepsilon S_g \rho_g\right)+\nabla \cdot\left(\rho_g \frac{K_g K_{rg}}{\mu_i}\nabla P\right)=R_{evap} \tag{1.76}$$

where the density is given by

$$\rho_g=\frac{PM_g}{RT} \tag{1.77}$$

where M_g is the molecular weight of the gas.

Using the same approach, Ni et al. (1999) analyzed moisture movement in wet biomaterials under intensive microwave heating (explained further in next section). The model used in the experiment included spatially varying and intense internal heat generation, evaporation and pressure-driven flow. The results suggested that transport is significantly enhanced by internal pressure gradients caused by internal heating and vaporization. A non-equilibrium multiphase model considering shrinkage and pore evolution for intermittent microwave-convective drying of food was developed by Joardder et al. (2017). The model proved its ability to manifest the characteristics of different parameters successfully, which is not possible using simpler models.

The conservation of energy in the Eulerian multiphase model is comprised of separate enthalpy equations for each phase:

Solid:

$$(1-\varepsilon)\frac{\partial}{\partial t}\left(\rho_s h_s\right)=\nabla \cdot\left((1-\varepsilon)k_s \nabla T_s\right)+h_{sl}A_{sl}\left(T_s-T_l\right)+h_{sg}A_{sg}\left(T_s-T_g\right) \tag{1.78}$$

Liquid:

$$\varepsilon\frac{\partial}{\partial t}\left(\alpha_l \rho_l h_l\right)+\nabla \cdot\left(\alpha_l \rho_l h_l \mathbf{u}_l\right)$$
$$=\nabla \cdot\left(\varepsilon\alpha_l k_l \nabla T_l\right)+h_{sl}A_{sl}\left(T_l-T_s\right)+h_{gl}A_{gl}\left(T_l-T_g\right) \tag{1.79}$$
$$+\left(Q_{gl}+\dot{m}_{gl}h_{gl}-\dot{m}_{lg}h_{lg}\right)$$

Gas:

$$\varepsilon\frac{\partial}{\partial t}\left(\alpha_g \rho_g h_g\right)+\nabla \cdot\left(\alpha_g \rho_g h_g \mathbf{u}_g\right)$$
$$=\nabla \cdot\left(\varepsilon\alpha_g k_g \nabla T_g\right)+h_{sg}A_{sg}\left(T_g-T_s\right) \tag{1.80}$$
$$+h_{lg}A_{lg}\left(T_g-T_l\right)+\left(Q_{lg}+\dot{m}_{lg}h_{lg}-\dot{m}_{gl}h_{gl}\right)$$

where h is the specific enthalpy of the phase, Q is the intensity of heat exchange between the liquid and gaseous phases, and h_{lg} is the interphase enthalpy (enthalpy of gas at temperature of the liquid, in the case of evaporation). The heat exchange between phases must be limited by additional condition: $Q_{lg} = -Q_{gl}$. The local thermal non-equilibrium (LTNE) heat transfer equations can be reduced into local thermal equilibrium (LTE) model similar to Eq. 1.37 provided the condition in Eqs. 1.20–1.24 are satisfied.

1.3.3   Conjugate Models

Thus far, modeling of the heat and mass transfer phenomena during drying has ignored effects of the environment of the porous materials. In reality, however, drying behavior is largely influenced by external parameters that in fact cannot be ignored. Therefore, it is appropriate to treat drying as a conjugate problem and take into account the effect of external conditions on the transport phenomena taking place within the material. The coupling of the porous medium model with the one for external flow is performed by the interface conditions (Erriguible et al. 2006). Murugesan et al. (2001) have carried out a two-dimensional analysis of brick drying, where Navier-Stokes equations including buoyancy terms were solved for the external flow field, along with conjugate heat and mass transfer in the brick. The rate of evaporation was found to be higher at the leading edge compared to other regions because of the thin concentration boundary layer. Kurnia et al. (2013, 2017) numerically investigated the performance of an impinging-jet drying system at various configurations using the CFD approach to solve the conjugated heat/mass transfer problem. The model considers slab material with drying chambers. Various factors affecting drying kinetics were investigated: substrate thickness, shapes, jet velocity, pulsation and intermittency. The results suggest that for thin slabs, impinging-jet with pulsating and intermittent flow offers comparable drying kinetics (as compared to that of a steady jet) with lower energy consumption and more uniform temperature distribution.

1.3.3.1   Convective Drying

Convective drying is the most common way of thermal drying. During convective drying of porous solids, transfer of energy from the surrounding environment and transfer of moisture from within the solid occur simultaneously (McMinn and Magee 1999). Modeling convective drying with the conjugate model allows accounting for spatial and temporal variations in convective boundary conditions and thereby circumvents the use of convective transfer coefficients (CTCs), which simplifies calculations. Defraeye et al. (2012) analyzed convective drying of an unsaturated porous flat plate at low Reynolds numbers (10^3) and determined the spatial and temporal variability of convective transfer coefficients via conjugate modeling. The drying behavior predicted by the conjugate model was compared with porous-material modeling using spatially and/or temporally variable CTCs. The drying rate was found to be lower with spatially variable CTCs due to the quick onset of the decreasing drying rate period. A temporal CTC variation was identified where distinct peaks in drying rate appeared at the surface right before the surface dried out locally, indicating that CTCs are strongly dependent on the temperature and moisture distribution in the boundary layer and at the air-porous material interface. Chandra Mohan and Talukdar (2010) developed a 3D heat and mass transfer model with consideration of variable heat transfer coefficient, variable diffusion coefficient and convective boundary condition to analyze the transient behavior of temperature and moisture content inside a moist object subjected to convective drying. The variable transfer coefficients required in the model were calculated using CFD simulations. The effects of air flow velocity and air temperature were evaluated, and it was found that increasing both parameters results in increase of both heat transfer coefficient and drying rate.

- Conservation equation of mass

$$\frac{\partial}{\partial t}(\rho) + \nabla \cdot (\rho \mathbf{U}) = 0 \tag{1.81}$$

The assumption of incompressible flow leads to:

$$\nabla \cdot \mathbf{U} = 0 \tag{1.82}$$

- Conservation equation of momentum

$$\frac{\partial}{\partial t}(\rho \mathbf{U}) + \nabla \cdot (\rho \mathbf{U}\mathbf{U}) = \nabla \cdot \left((\mu + \mu_t)(\nabla \mathbf{U} + \nabla \mathbf{U}^T) \right) - \nabla P + \rho \vec{g} \tag{1.83}$$

where the velocity $\mathbf{U}$ is comprised of mean velocity $\bar{U}$ and fluctuating velocity due to turbulence U'

$$\mathbf{U} = \bar{U} + U' \tag{1.84}$$

- Conservation equation of energy
  The turbulence energy equation can be written in terms of enthalpy to be consistent with the one associated with the porous medium, namely,

$$\frac{\partial}{\partial t}(\rho h) + \nabla \cdot (\rho h \mathbf{u}) = \nabla \cdot \left[\left(k + \frac{c_p \mu_t}{\mathrm{Pr}_t} \right) \nabla T \right] \tag{1.85}$$

where the right-hand side of the equation represents the turbulent diffusion.
- Turbulence Model
  There are many turbulent models available in literature, such as k-epsilon, k-omega, Reynolds Stress Model (RSM) and so forth. One of the most-used turbulence models in drying is the k-epsilon model. The model considers a two-equation model that solves for turbulent kinetic

energy κ and rate of dissipation ϵ. The equations for turbulent kinetic energy is given by:

$$\frac{\partial}{\partial t}(\rho\kappa) + \nabla \cdot (\rho\kappa\mathbf{U}) = \nabla \cdot \left[\left(\mu + \frac{\mu_t}{\sigma_\kappa}\right)\nabla\kappa\right] + G_\kappa - \rho\epsilon \qquad (1.86)$$

and its rate of dissipation is:

$$\frac{\partial}{\partial t}(\rho\epsilon) + \nabla \cdot (\rho\epsilon\mathbf{U}) = \nabla \cdot \left[\left(\mu + \frac{\mu_t}{\sigma_\epsilon}\right)\nabla\varepsilon\right] + C_{1\epsilon}G_\kappa\frac{\epsilon}{\kappa} - C_{2\epsilon}\rho\frac{\epsilon^2}{\kappa} \qquad (1.87)$$

where σ_κ and σ_ϵ are the turbulent Prandtl numbers for κ and ϵ, respectively, G_κ represents the generation of turbulence energy due to the mean velocity gradients and $C_{1\epsilon}$, $C_{2\epsilon}$ are constants. The turbulent viscosity μ_t is defined in terms of κ and ϵ, as follows:

$$\mu_t = \rho C_\mu \frac{k^2}{\epsilon} \qquad (1.88)$$

where C_μ is constant.

In the conjugate convective drying model, special attention has to be taken into account at the interface between drying substrate and the environment. The continuation of mass, momentum and energy equations between porous media (drying substrate) and plain media (ambient) should be established to allow for conjugate transfer.

1.3.3.2 Microwave Drying

Microwave drying is extensively employed in porous media drying due to its rapid drying rate and uniform drying, which results from volumetric heating and internal evaporation (Zhou et al. 2018). Typically, microwaves are used to assist or enhance another drying operation, resulting in combined drying processes, such as microwave-convective drying (Kowalski et al. 2010; Turner and Jolly 1991; Sharma and Prasad 2004), microwave-assisted vacuum drying (Drouzas and Schubert 1996; Monteiro et al. 2018; Song et al. 2018), hot air–microwave drying (Maskan 2001; Andrés et al. 2004; Wang et al. 2018) and microwave freeze drying (Duan et al. 2010; Ambros et al. 2018; Abbasi and Azari 2009). It has been proven that microwave drying or microwave-assisted heating may considerably increase drying rate and produce better-quality dried products when compared to the drying processes unaccompanied by microwaves. For example, Maskan (2000) compared drying characteristics, color and rehydration of banana samples dried by hot air, microwave and hot air followed by microwave finished drying. While it was found to have little effect on the color and rehydration capacity of finished products, microwave finish drying increased the drying rate and reduced drying time significantly.

The microwave drying of a multiphase porous medium involves electromagnetic heating, evaporation phase change, fluid flow and heat and mass transfer (Zhou et al. 2018). The electric field distribution is identified by Maxwell's equations for electromagnetics:

$$\nabla \cdot D = \rho \tag{1.89}$$

$$\nabla \cdot B = 0 \tag{1.90}$$

$$\frac{\partial B(u,t)}{\partial t} = -\nabla \times E(u,t) \tag{1.91}$$

$$\frac{\partial D(u,t)}{\partial t} = \nabla \times H(u,t) - J_e(u,t) \tag{1.92}$$

where D is the electric displacement, ρ is electric charge density, B is the magnetic flux density, E is electric field intensity, H represents the magnetic field and J denotes the electric current density. For linear, isotropic and non-dispersive material:

$$D = \varepsilon E \tag{1.93}$$

$$B = \mu H \tag{1.94}$$

$$J = \sigma E \tag{1.95}$$

where ε is the electric permittivity, μ is the magnetic permeability and σ is the electrical conductivity.

When the material absorbs microwave energy over time, the temperature difference causes the heat to be conducted throughout the volume of the material (Hansson and Anttii 2008). The energy balance equation to describe heat transfer could be any of the energy balances mentioned in previous sections, depending on the model used, with an additional heat source term S_{Hm} due to internal heating from the dielectric properties from microwave irradiation. The heat source, also called 'electromagnetic losses', consists of two parts: resistive heating losses (ohmic heating) due to electric current Q_{rh} and magnetic losses Q_{ml}, given in Eqs. 1.97–1.98, respectively.

$$S_{Hm} = Q_{rh} + Q_{ml} \tag{1.96}$$

$$Q_{rh} = 2\pi f \, \varepsilon_0 \varepsilon_r E^2 \tag{1.97}$$

$$Q_{ml} = \frac{1}{2} \omega \mu_0 \mu_r |H|^2 \tag{1.98}$$

where f is frequency, ε_0 is the permittivity of free space, ε_r is the relative dielectric loss factor, ω is the angular frequency, μ_0 is the permeability of free space and μ_r is the relative magnetic loss factor.

1.3.3.3 Electrohydrodynamic Drying

Over the last few decades, the enhancement of heat and mass transfer by electric field has been a subject of great interest. Wolny and Kaniuk (1996) studied the effect of an electric field on evaporation of water of a wet kaolin sample placed in a flat grounded box and found that an electric field that generates an ionic wind can significantly intensify heat and mass transfer in drying processes with up to eight times higher mean values of heat and mass transfer coefficients. Allen and Karayiannis (1995) have reviewed published work on Electrohydrodynamic (EHD)-enhanced single-phase and two-phase heat transfer and concluded that single-phase heat transfer rates can be significantly enhanced, with the highest enhancement ratio given by flow of 'corona (electric) wind' and by electrophoresis. EHD dehydration is a relatively new alternative technique used to dry porous solids that allows lower energy consumption. EHD drying has been used to dry various materials, like sand (Singh et al. 2017), apple (Hashinaga et al. 1999), potato slabs (Chen et al. 1994), wheat (Cao et al. 2004), tomato (Esehaghbeygi and Basiry 2011), and partially wetted glass beads (Alem-Rajabif and Lai 2005). It utilizes the secondary flow generated by high-voltage corona discharge, which induces ionic wind (Defraeye and Martynenko 2018). Zhao and Adamiak (2005) have numerically and experimentally investigated the EHD flow in air produced by electric corona discharge in the pin-plane and pin-mesh configurations. The study concluded that airflow velocity profiles and the pressure distribution are greatly affected by the applied voltage and the corona device configuration. Electrohydrodynamic drying is a non-thermal drying, which makes it suitable for dehydrating heat-sensitive biomaterials and perishable food. Literature reviews and prospects of EHD drying have been examined by Bajgai et al. (2006) and Defraeye and Martynenko (2018), among other authors.

1.4 CONCLUDING REMARKS

This chapter presents an overview of the recent developments in mathematical modeling of drying in porous media. Various drying models have been reported in the literature, from simple liquid diffusion model to complex multi-fluid multiphase models and the conjugate model. Although the use of drying model depends on the purpose and practicality of the model, a complex model that was prohibitive computationally in the past could be feasible using current computing power. The development of the mathematical model itself faces several challenges which need to be addressed:

- The development of accurate interphase mass, momentum and energy in multi-fluid drying model needs further attention to take into account liquid evaporation, drag and drying rate both in the representative elementary volume and the pore scale level.

- Model validation typically refers only to the global drying rate. More thorough validation needs three-dimensional local distributions of temperature and liquid fraction over time.
- Multimode conjugate drying models with more than one drying mechanism, such as convective-conductive-radiative-microwave drying, need to be developed and validated as many innovative drying approaches include use combined modes of energy input sequentially or concurrently.

REFERENCES

Abbasi, S., and S. Azari. 2009. Novel Microwave–Freeze Drying of Onion Slices. *International Journal of Food Science & Technology* 44, no. 5 (May 1): 974–979.

Adamski, R., and Z. Pakowski. 2013. Identification of Effective Diffusivities in Anisotropic Material of Pine Wood during Drying with Superheated Steam. *Drying Technology* 31, no. 3 (February 17): 264–268.

Alem-Rajabif, A., and F.C. Lai. 2005. EHD-Enhanced Drying of Partially Wetted Glass Beads. *Drying Technology* 23, no. 3 (March 30): 597–609.

Allen, P.H.G., and T.G. Karayiannis. 1995. Electrohydrodynamic Enhancement of Heat Transfer and Fluid Flow. *Heat Recovery Systems and CHP* 15, no. 5 (July 1): 389–423.

Ambros, S., R. Mayer, B. Schumann, and U. Kulozik. 2018. Microwave-Freeze Drying of Lactic Acid Bacteria: Influence of Process Parameters on Drying Behavior and Viability. *Innovative Food Science & Emerging Technologies* 48 (August 1): 90–98.

Andrés, A., C. Bilbao, and P. Fito. 2004. Drying Kinetics of Apple Cylinders under Combined Hot Air–Microwave Dehydration. *Journal of Food Engineering* 63, no. 1 (June 1): 71–78.

Auton, T.R. 1987. The Lift Force on a Spherical Body in a Rotational Flow. *Journal of Fluid Mechanics* 183 (October): 199–218.

Azzouz, S., K.B. Dhib, R. Bahar, S. Ouertani, M.T. Elaieb, and A. Elcafsi. 2018. Mass Diffusivity of Different Species of Wood in Convective Drying. *European Journal of Wood and Wood Products* 76, no. 2 (March 1): 573–582.

Azzouz, S., A. Guizani, W. Jomaa, and A. Belghith. 2002. Moisture Diffusivity and Drying Kinetic Equation of Convective Drying of Grapes. *Journal of Food Engineering* 55, no. 4 (December 1): 323–330.

Bajgai, T.R., G.S.V. Raghavan, F. Hashinaga, and M.O. Ngadi. 2006. Electrohydrodynamic Drying—A Concise Overview. *Drying Technology* 24, no. 7 (August 1): 905–910.

Bažant, Z.P., and L.J. Najjar. 1971. Drying of Concrete as a Nonlinear Diffusion Problem. *Cement and Concrete Research* 1, no. 5 (September 1): 461–473.

Beavers, G.S., E.M. Sparrow, and D.E. Rodenz. 1973. Influence of Bed Size on the Flow Characteristics and Porosity of Randomly Packed Beds of Spheres. *Journal of Applied Mechanics* 40, no. 3 (September 1): 655–660.

Bird, R.B. 2007. *Transport Phenomena*. New York : J. Wiley,.

Bosl, W.J., J. Dvorkin, and A. Nur. 1998. A Study of Porosity and Permeability Using a Lattice Boltzmann Simulation. *Geophysical Research Letters* 25, no. 9 (May 1): 1475–1478.

Bowen, R.M. 1976. Part I - Theory of Mixtures. In *Continuum Physics*, ed. A.C. Eringen, 1–127. Academic Press. http://www.sciencedirect.com/science/article/pii/B9780122408038500177

Brackbill, J.U., D.B. Kothe, and C. Zemach. 1992. A Continuum Method for Modeling Surface Tension. *Journal of Computational Physics* 100, no. 2 (June 1): 335–354.

Bronlund, J. 1997. The Modeling of Caking in Bulk Lactose. Thesis, Massey University. https://mro.massey.ac.nz/handle/10179/2738

Bruce Litchfield, J., and M.R. Okos. 1992. Moisture Diffusivity in Pasta during Drying. *Journal of Food Engineering* 17, no. 2 (January 1): 117–142.

Bryant, S., and M. Blunt. 1992. Prediction of Relative Permeability in Simple Porous Media. *Physical Review A* 46, no. 4 (August 1): 2004–2011.

Cao, W., Y. Nishiyama, and S. Koide. 2004. Electrohydrodynamic Drying Characteristics of Wheat Using High Voltage Electrostatic Field. *Journal of Food Engineering* 62, no. 3 (May 1): 209–213.

Chandra Mohan, V.P., and P. Talukdar. 2010. Three Dimensional Numerical Modeling of Simultaneous Heat and Moisture Transfer in a Moist Object Subjected to Convective Drying. *International Journal of Heat and Mass Transfer* 53, no. 21–22 (October): 4638–4650.

Chemkhi, S., W. Jomaa, and F. Zagrouba. 2009. Application of a Coupled Thermo-Hydro-Mechanical Model to Simulate the Drying of Nonsaturated Porous Media. *Drying Technology* 27, no. 7–8 (July 29): 842–850.

Chen, D., Y. Zhang, and X. Zhu. 2012. Drying Kinetics of Rice Straw under Isothermal and Nonisothermal Conditions: A Comparative Study by Thermogravimetric Analysis. *Energy & Fuels* 26, no. 7 (July 19): 4189–4194.

Chen, D., Y. Zheng, and X. Zhu. 2012. Determination of Effective Moisture Diffusivity and Drying Kinetics for Poplar Sawdust by Thermogravimetric Analysis under Isothermal Condition. *Bioresource Technology* 107 (March 1): 451–455.

Chen, Y., N.N. Barthakur, and N.P. Arnold. 1994. Electrohydrodynamic (EHD) Drying of Potato Slabs. *Journal of Food Engineering* 23, no. 1 (January 1): 107–119.

Chikh, S., A. Boumedien, K. Bouhadef, and G. Lauriat. 1995. Non-Darcian Forced Convection Analysis in an Annulus Partially Filled with a Porous Material. *Numerical Heat Transfer, Part A: Applications* 28, no. 6 (December): 707–722.

Corey, A. T. 1954. The interrelation between gas and oil relative permeabilities. *Producers Monthly* 19, 38.

Defraeye, T., B. Blocken, and J. Carmeliet. 2012. Analysis of Convective Heat and Mass Transfer Coefficients for Convective Drying of a Porous Flat Plate by Conjugate Modeling. *International Journal of Heat and Mass Transfer* 55, no. 1–3 (January 15): 112–124.

Defraeye, T., and A. Martynenko. 2018. Future Perspectives for Electrohydrodynamic Drying of Biomaterials. *Drying Technology* 36, no. 1 (January 2): 1–10.

Dincer, I., and S. Dost. 1995. An Analytical Model for Moisture Diffusion in Solid Objects during Drying. *Drying Technology* 13, no. 1–2 (January 1): 425–435.

Dolinskiy, A.A., A.S. Dorfman, and B.V. Davydenko. 1991. Conjugate Heat and Mass Transfer in Continuous Processes of Convective Drying. *International Journal of Heat and Mass Transfer* 34, no. 11 (November 1): 2883–2889.

Doymaz, İ. 2005. Drying Characteristics and Kinetics of Okra. *Journal of Food Engineering* 69, no. 3 (August 1): 275–279.

Drew, D.A., and R.T. Lahey. 1987. The Virtual Mass and Lift Force on a Sphere in Rotating and Straining Inviscid Flow. *International Journal of Multiphase Flow* 13, no. 1 (January 1): 113–121.

Drew, D.A., and R.T. Lahey. 1990. Some Supplemental Analysis Concerning the Virtual Mass and Lift Force on a Sphere in a Rotating and Straining Flow. *International Journal of Multiphase Flow* 16, no. 6 (November 1): 1127–1130.

Drouzas, A.E., and H. Schubert. 1996. Microwave Application in Vacuum Drying of Fruits. *Journal of Food Engineering* 28, no. 2 (May 1): 203–209.

Duan, X., M. Zhang, A.S. Mujumdar, and S. Wang. 2010. Microwave Freeze Drying of Sea Cucumber (Stichopus Japonicus). *Journal of Food Engineering* 96, no. 4 (February 1): 491–497.

Ece, M.C., and A. Cihan. 1993. A Liquid Diffusion Model for Drying Rough Rice. *Transactions of the ASAE* 36, no. 3: 837–840.

El-Batsh, H.M., M.A. Doheim, and A.F. Hassan. 2012. On the Application of Mixture Model for Two-Phase Flow Induced Corrosion in a Complex Pipeline Configuration. *Applied Mathematical Modeling* 36, no. 11 (November 1): 5686–5699.

Elbert, G., M.P. Tolaba, R.J. Aguerre, and C. Suárez. 2001. A Diffusion Model with a Moisture Dependent Diffusion Coefficient for Parboiled Rice. *Drying Technology* 19, no. 1 (January 31): 155–166.

Erriguible, A., P. Bernada, F. Couture, and M. Roques. 2006. Simulation of Convective Drying of a Porous Medium with Boundary Conditions Provided by CFD. *Chemical Engineering Research and Design* 84, no. 2 (February 1): 113–123.

Esehaghbeygi, A., and M. Basiry. 2011. Electrohydrodynamic (EHD) Drying of Tomato Slices (Lycopersicon Esculentum). *Journal of Food Engineering* 104, no. 4 (June 1): 628–631.

Fan, C.C., S.P. Liaw, W.R. Fu, and B.S. Pan. 2003. Mathematical Model for Prediction of Intermittent Drying and Pressing Process of Mullet Roe. *Journal of Food Science* 68, no. 3 (April 1): 886–891.

Feng, H., J. Tang, R.P. Cavalieri, and O.A. Plumb. 2001. Heat and Mass Transport in Microwave Drying of Porous Materials in a Spouted Bed. *AIChE Journal* 47, no. 7 (July 1): 1499–1512.

Franco, C.M.R., A.G.B. da Lima, J.V. Silva, and A.G. Nunes. 2016. Applying Liquid Diffusion Model for Continuous Drying of Rough Rice in Fixed Bed. *Defect and Diffusion Forum; Zurich* 369 (July): 152–156.

Gandomkar, A., and K.E. Gray. 2018. Local Thermal Non-Equilibrium in Porous Media with Heat Conduction. *International Journal of Heat and Mass Transfer* 124 (September 1): 1212–1216.

Hansson, L., and L. Antti. 2008. Modeling Microwave Heating and Moisture Redistribution in Wood. *Drying Technology* 26, no. 5 (April 7): 552–559.

Hashinaga, F., T.R. Bajgai, S. Isobe, and N.N. Barthakur. 1999. Electrohydrodynamic (EHD) Drying of Apple Slices. *Drying Technology* 17, no. 3 (March 1): 479–495.

Hassini, L., S. Azzouz, R. Peczalski, and A. Belghith. 2007. Estimation of Potato Moisture Diffusivity from Convective Drying Kinetics with Correction for Shrinkage. *Journal of Food Engineering* 79, no. 1 (March 1): 47–56.

Hii, C.L., C.L. Law, and M. Cloke. 2009. Modeling Using a New Thin Layer Drying Model and Product Quality of Cocoa. *Journal of Food Engineering* 90, no. 2 (January): 191–198.

Hirsch, C. 2007. *Numerical Computation of Internal and External Flows : Fundamentals of Computational Fluid Dynamics*. Amsterdam: Elsevier/Butterworth-Heinemann.

Jang, J.Y., and J.L. Chen. 1992. Forced Convection in a Parallel Plate Channel Partially Filled with a High Porosity Medium. *International Communications in Heat and Mass Transfer* 19, no. 2 (March 1): 263–273.

Joardder, M.U.H., C. Kumar, and M.A. Karim. 2017. Multiphase Transfer Model for Intermittent Microwave-Convective Drying of Food: Considering Shrinkage and Pore Evolution. *International Journal of Multiphase Flow* 95 (October 1): 101–119.

Kamyabi, F. 2014. *Multiphase Flow in Porous Media*. LAP LAMBERT Academic Publishing. https://www.morebooks.de/store/gb/book/multiphase-flow-in-porous-media/isbn/978-3-659-61506-1

Katekawa, M.E., and M.A. Silva. 2006. A Review of Drying Models Including Shrinkage Effects. *Drying Technology* 24, no. 1 (February 1): 5–20.

Kaviany, M. 1999. *Principles of Heat Transfer in Porous Media*. New York : Springer-Verlag.

Kowalski, S.J., G. Musielak, and J. Banaszak. 2010. Heat and Mass Transfer during Microwave-Convective Drying. *AIChE Journal* 56, no. 1 (January 1): 24–35.

Kumar, C., M.U.H. Joardder, T.W. Farrell, and M.A. Karim. 2016. Multiphase Porous Media Model for Intermittent Microwave Convective Drying (IMCD) of Food. *International Journal of Thermal Sciences* 104 (June 1): 304–314.

Kumar, Chandan, M.U.H. Joardder, T.W. Farrell, G.J. Millar, and A. Karim. 2014. Multiphase Porous Media Model for Heat and Mass Transfer during Drying of Agricultural Products. In *Science & Engineering Faculty*. RMIT University, Melbourne, VIC. http://www.afms.org.au/openconf/modules/request.php?module=oc_proceedings&action=view.php&a=Accept&id=466

Kurnia, J.C., A.P. Sasmito, W. Tong, and A.S. Mujumdar. 2013. Energy-Efficient Thermal Drying Using Impinging-Jets with Time-Varying Heat Input – A Computational Study. *Journal of Food Engineering* 114, no. 2 (January 1): 269–277.

Kurnia, J.C., A.P. Sasmito, P. Xu, and A.S. Mujumdar. 2017. Performance and Potential Energy Saving of Thermal Dryer with Intermittent Impinging Jet. *Applied Thermal Engineering* 113 (February 25): 246–258.

Lemus-Mondaca, R.A., C.E. Zambra, A. Vega-Gálvez, and N.O. Moraga. 2013. Coupled 3D Heat and Mass Transfer Model for Numerical Analysis of Drying Process in Papaya Slices. *Journal of Food Engineering* 116, no. 1 (May 1): 109–117.

Li, K. 2008. A New Method for Calculating Two-Phase Relative Permeability from Resistivity Data in Porous Media. *Transport in Porous Media* 74, no. 1 (August 1): 21–33.

Li, K., and R.N. Horne. 2006. Comparison of Methods to Calculate Relative Permeability from Capillary Pressure in Consolidated Water-Wet Porous Media. *Water Resources Research* 42, no. 6 (June 1). https://agupubs.onlinelibrary.wiley.com/doi/abs/10.1029/2005WR004482

Li, Z., and N. Kobayashi. 2005. Determination of Moisture Diffusivity by Thermo-Gravimetric Analysis under Non-Isothermal Condition. *Drying Technology* 23, no. 6 (June): 1331–1342.

López-Méndez, E.M., B. Ortiz-García-Carrasco, H. Ruiz-Espinosa, A. Sampieri-Croda, M.A. García-Alvarado, C.E. Ochoa-Velasco, A. Escobedo-Morales, and I.I. Ruiz-López. 2018. Effect of Shape Change and Initial Geometry on Water Diffusivity Estimation during Drying of Gel Model Systems. *Journal of Food Engineering* 216 (January 1): 52–64.

Manninen, M., V. Taivassalo, and S. Kallio. 1996. *On the Mixture Model for Multiphase Flow*. VTT Publications 288. Espoo: Technical Research Centre of Finland.

Markicevic, B., and N. Djilali. 2006. Two-Scale Modeling in Porous Media: Relative Permeability Predictions. *Physics of Fluids* 18, no. 3 (March 1): 033101.

Maskan, M. 2000. Microwave/Air and Microwave Finish Drying of Banana. *Journal of Food Engineering* 44, no. 2 (May 1): 71–78.

Maskan 2001. Drying, Shrinkage and Rehydration Characteristics of Kiwifruits during Hot Air and Microwave Drying. *Journal of Food Engineering* 48, no. 2 (May 1): 177–182.

McMinn, W.A.M., and T.R. Magee. 1999. Principles, Methods and Applications of the Convective Drying of Foodstuffs. *Food and Bioproducts Processing* 77, no. 3 (September 1): 175–193.

Mihoubi, D., F. Zagrouba, J. Vaxelaire, A. Bellagi, and M. Roques. 2004. Transfer Phenomena During the Drying of a Shrinkable Product: Modeling and Simulation. *Drying Technology* 22, no. 1–2 (December 31): 91–109.

Minkowycz, W.J., A. Haji-Sheikh, and K. Vafai. 1999. On Departure from Local Thermal Equilibrium in Porous Media Due to a Rapidly Changing Heat Source: The Sparrow Number. *International Journal of Heat and Mass Transfer* 42, no. 18 (September 1): 3373–3385.

Mohamad, A.A. 2003. Heat Transfer Enhancements in Heat Exchangers Fitted with Porous Media Part I: Constant Wall Temperature. *International Journal of Thermal Sciences* 42, no. 4 (April 1): 385–395.

Monteiro, R.L., J.V. Link, G. Tribuzi, B.A.M. Carciofi, and J.B. Laurindo. 2018. Microwave Vacuum Drying and Multi-Flash Drying of Pumpkin Slices. *Journal of Food Engineering* 232 (September 1): 1–10.

Moraga, F., Bonetto, F.J., and R. Lahey. 1999. Lateral Forces on Spheres in Turbulent Uniform Shear Flow. *International Journal of Multiphase Flow* 25 (September 1): 1321–1372.

Morsi, S.A., and A.J. Alexander. 1972. An Investigation of Particle Trajectories in Two-Phase Flow Systems. *Journal of Fluid Mechanics* 55, no. 2 (September): 193–208.

Murugesan, K., H.N. Suresh, K.N. Seetharamu, P.A. Aswatha Narayana, and T. Sundararajan. 2001. A Theoretical Model of Brick Drying as a Conjugate Problem. *International Journal of Heat and Mass Transfer* 44, no. 21 (November 1): 4075–4086.

Muskat, M., and M.W. Meres. 1936. The Flow of Heterogeneous Fluids through Porous Media. *Physics* 7 (September 1): 346–363.

Muzaferija, S., M. Peric, P. Sames, and T. Schellin. 1999. *A Two-Fluid Navier-Stokes Solver to Simulate Water Entry in Symposium on Naval Hydrodynamics*, 638–651. Washington, DC: National Academy Press.

Nadler, K.C., E.T. Choong, and D.M. Wetzel. 1985. Mathematical Modeling of the Diffusion of Water in Wood During Drying. *Wood and Fiber Science* 17, no. 3 (June 27): 404–423.

Ni, H., A.K. Datta, and K.E. Torrance. 1999. Moisture Transport in Intensive Microwave Heating of Biomaterials: A Multiphase Porous Media Model. *International Journal of Heat and Mass Transfer* 42, no. 8 (April 1): 1501–1512.

Niamnuy, C., S. Devahastin, S. Soponronnarit, and G.S. Vijaya Raghavan. 2008. Modeling Coupled Transport Phenomena and Mechanical Deformation of Shrimp during Drying in a Jet Spouted Bed Dryer. *Chemical Engineering Science* 63, no. 22 (November 1): 5503–5512.

Nield, D.A., and A. Bejan. 2017. *Convection in Porous Media*. 5th ed. Springer International Publishing. www.springer.com/gp/book/9783319495613

Özdemir, M., and Y. Onur Devres. 1999. The Thin Layer Drying Characteristics of Hazelnuts during Roasting. *Journal of Food Engineering* 42, no. 4 (December 1): 225–233.

Pacheco-Aguirre, F.M., A. Ladrón-González, H. Ruiz-Espinosa, M.A. García-Alvarado, and I.I. Ruiz-López. 2014. A Method to Estimate Anisotropic Diffusion Coefficients for Cylindrical Solids: Application to the Drying of Carrot. *Journal of Food Engineering* 125 (March 1): 24–33.

Park, K.J. 1998. Diffusional Model with and without Shrinkage during Salted Fish Muscle Drying. *Drying Technology* 16, no. 3–5 (January): 889–905.

Parti, M., and I. Dugmanics. 1990. Diffusion Coefficient for Corn Drying. *Transactions of the ASAE* 33, no. 5: 1652.

Plumb, O.A., G.A. Spolek, and B.A. Olmstead. 1985. Heat and Mass Transfer in Wood during Drying. *International Journal of Heat and Mass Transfer* 28, no. 9 (September 1): 1669–1678.

Purcell, W.R. 1949. Capillary Pressures - Their Measurement Using Mercury and the Calculation of Permeability Therefrom. *Journal of Petroleum Technology* 1, no. 02 (February 1): 39–48.

Resende, O., P.C. Corrêa, C. Jarén, and A.J. Moure. 2007. Bean Moisture Diffusivity and Drying Kinetics: A Comparison of the Liquid Diffusion Model When Taking into Account and Neglecting Grain Shrinkage. *Spanish Journal of Agricultural Research* 5, no. 1 (March 1): 51.

Rohsenow, W.M., J.P. Hartnett, and Y.I. Cho. 1998. *Handbook of Heat Transfer (3rd Edition)*. New York, USA: McGraw Hill Professional Publishing. http://ebookcentral.proquest.com/lib/mcgill/detail.action?docID=4657093

Rosselló, C., S. Simal, N. SanJuan, and A. Mulet. 1997. Nonisotropic Mass Transfer Model for Green Bean Drying. *Journal of Agricultural and Food Chemistry* 45, no. 2 (February): 337–342.

Schiller, L., and A. Naumann. 1935. A Drag Coefficient Correlation. *Z. Ver. Deutsch. Ing* 77: 318–320. Scopus.

Sen, N., K.K. Singh, A.W. Patwardhan, S. Mukhopadhyay, and K.T. Shenoy. 2016. CFD Simulation of Two-Phase Flow in Pulsed Sieve-Plate Column – Identification of a Suitable Drag Model to Predict Dispersed Phase Hold Up. *Separation Science and Technology* 51, no. 17 (November 21): 2790–2803.

Sharma, G.P., and S. Prasad. 2004. Effective Moisture Diffusivity of Garlic Cloves Undergoing Microwave-Convective Drying. *Journal of Food Engineering* 65, no. 4 (December 1): 609–617.

Sharma, S., and D.A. Siginer. 2010. Permeability Measurement Methods in Porous Media of Fiber Reinforced Composites. *Applied Mechanics Reviews* 63, no. 2 (February 22): 020802-020802–19.

Sherwood, T.K. 1929. The Drying of Solids—I. *Industrial & Engineering Chemistry* 21, no. 1 (January): 12–16.

da Silva, W.P., I. Hamawand, and C.M.D.P.S. e Silva. 2014. A Liquid Diffusion Model to Describe Drying of Whole Bananas Using Boundary-Fitted Coordinates. *Journal of Food Engineering* 137 (September 1): 32–38.

Silva, W.P. da, C.M.D.P. da Silva e Silva, J.W. Precker, J.P. Gomes, P.L. Nascimento, and L.D. da Silva. 2012. Diffusion Models for the Description of Seedless Grape Drying Using Analytical and Numerical Solutions. *Agricultural Sciences* 3, no. 4: 545–556.

Simal, S., A. Mulet, J. Tarrazo, and C. Rosselló. 1996. Drying Models for Green Peas. *Food Chemistry* 55, no. 2 (January 1): 121–128.

Singh, A., S.K.K. Vanga, G.R. Nair, Y. Gariepy, V. Orsat, and V. Raghavan. 2017. Electrohydrodynamic Drying of Sand. *Drying Technology* 35, no. 3 (February 17): 312–322.

Sobieski, W. 2008. Numerical Analysis of Sensitivity of Eulerian Multiphase Model for a Spouted-Bed Grain Dryer. *Drying Technology* 26, no. 12 (November 21): 1438–1456.

Song, C., T. Wu, Z. Li, J. Li, and H. Chen. 2018. Analysis of the Heat Transfer Characteristics of Blackberries during Microwave Vacuum Heating. *Journal of Food Engineering* 223 (April 1): 70–78.

Souraki, B.A., and D. Mowla. 2008. Axial and Radial Moisture Diffusivity in Cylindrical Fresh Green Beans in a Fluidized Bed Dryer with Energy Carrier: Modeling with and without Shrinkage. *Journal of Food Engineering* 88, no. 1 (September 1): 9–19.

Srikiatden, J., and J.S. Roberts. 2006. Measuring Moisture Diffusivity of Potato and Carrot (Core and Cortex) during Convective Hot Air and Isothermal Drying. *Journal of Food Engineering* 74, no. 1 (May 1): 143–152.

Sun, L., M.R. Islam, J.C. Ho, and A.S. Mujumdar. 2005. A Diffusion Model for Drying of a Heat Sensitive Solid under Multiple Heat Input Modes. *Bioresource Technology* 96, no. 14 (September 1): 1551–1560.

Teixeira, M.B.F., and S. Tobinaga. 1998. A Diffusion Model for Describing Water Transport in Round Squid Mantle during Drying with a Moisture-Dependent Effective Diffusivity. *Journal of Food Engineering* 36, no. 2 (May 1): 169–181.

Treybal, R.E. 1968. *Mass Transfer Operations*. New York: McGraw Hill.

Turner, I., and A.S. Mujumdar. 1996. *Mathematical Modeling and Numerical Techniques in Drying Technology*. CRC Press. www.taylorfrancis.com/books/9781482292190

Turner, I.W., and P.C. Jolly. 1991. Combined Microwave and Convective Drying of a Porous Material. *Drying Technology* 9, no. 5 (December 1): 1209–1269.

Turner, I.W., and P. Perré. 2004. Vacuum Drying of Wood with Radiative Heating: II. Comparison between Theory and Experiment. *AIChE Journal* 50, no. 1 (January 1): 108–118.

Ubbink, O. 1997. Numerical Prediction of Two Fluid Systems with Sharp Interfaces. Ph.D., Imperial College London (University of London). http://hdl.handle.net/10044/1/8604

Waananen, K.M., and M.R. Okos. 1996. Effect of Porosity on Moisture Diffusion during Drying of Pasta. *Journal of Food Engineering* 28, no. 2 (May 1): 121–137.

Wang, Y., X. Li, X. Chen, B. Li, X. Mao, J. Miao, C. Zhao, L. Huang, and W. Gao. 2018. Effects of Hot Air and Microwave-Assisted Drying on Drying Kinetics, Physicochemical Properties, and Energy Consumption of Chrysanthemum. *Chemical Engineering and Processing - Process Intensification* 129 (July 1): 84–94.

Werner, S.R.L., R.L. Edmonds, J.R. Jones, J.E. Bronlund, and A.H.J. Paterson. 2008. Single Droplet Drying: Transition from the Effective Diffusion Model to a Modified Receding Interface Model. *Powder Technology* 179, no. 3. WCPT5 (January 1): 184–189.

Wolny, A., and R. Kaniuk. 1996. The Effect of Electric Field on Heat and Mass Transfer. *Drying Technology* 14, no. 2 (January 1): 195–216.

Yazbek, W., P. Pré, and A. Delebarre. 2006. Adsorption and Desorption of Volatile Organic Compounds in Fluidized Bed. *Journal of Environmental Engineering-Asce - J ENVIRON ENG-ASCE* 132 (May 1).

Yuan, Y., J. Zhang, D. Wang, Y. Xu, and B. Bhandari. 2018. Molecular Dynamics Simulation on Moisture Diffusion Process for Drying of Porous Media in Nanopores. *International Journal of Heat and Mass Transfer* 121 (June 1): 555–564.

Zhao, L., and K. Adamiak. 2005. EHD Flow in Air Produced by Electric Corona Discharge in Pin–Plate Configuration. *Journal of Electrostatics* 63, no. 3. Selected Papers from the ESA 2004 Annual Conference (March 1): 337–350.

Zhou, J., X. Yang, H. Zhu, J. Yuan, and K. Huang. 2018. Microwave Drying Process of Corns Based on Double-Porous Model. *Drying Technology* in press, doi: 10.1080/07373937.2018.1439952 (February 27): 1–13.

Zielinska, M., and M. Markowski. 2007. Drying Behavior of Carrots Dried in a Spout–Fluidized Bed Dryer. *Drying Technology* 25, no. 1 (February 12): 261–270.

2 Diffusivity in Drying of Porous Media

Chien Hwa Chong
Heriot-Watt University Malaysia Campus,
Precinct 5, Putrajaya, Malaysia

Chung Lim Law
University of Nottingham Malaysia Campus,
Semenyih, Selangor Darul Ehsan, Malaysia

Adam Figiel
Wroclaw University of Environmental and
Life Sciences, 51-630 Wrocław, Poland

Tommy Asni
Heriot-Watt University Malaysia Campus,
Precinct 5, Putrajaya, Malaysia

CONTENTS

2.1 INTRODUCTION TO DRYING TECHNOLOGY AND POROUS MEDIA

2.1.1 DRYING TECHNOLOGY

Drying is the process of removing moisture from a solid, semi-solid or liquid that involves simultaneous heat and mass transfer. Heat is transferred to porous media by a combination of conduction, convection and radiation depending on the type of drying process. The moisture can be evaporated from the surface (free moisture) or be transferred from within the body of porous solid to the surface (bound moisture). The speed at which the moisture is removed is known as the drying rate. Drying rate is affected by many factors; one of the factors that affects the drying rate is the material properties. In this case, the complex structures of porous media play a major role in determining the drying rates.

The drying period in a plot graphing drying rate versus moisture content is normally divided into three distinct periods; namely, initial transient period, constant rate period and falling rate period (Figure 2.1). The material heats up and the temperature rises to a certain degree during the transient period. At the constant rate period, a layer of free water normally covers the material; the mass-transport mechanism is evaporation. Therefore, the removal rate of water from the wet material is constant. In this period, diffusion is not significant, therefore determination of the diffusion coefficient is practically impossible as the excess of surface water is removed at the constant drying rate, which depends on external conditions such as velocity, humidity and temperature of the drying agent. This transfer process is governed by the external heat and mass transfer rate. After the surface of the material is partially

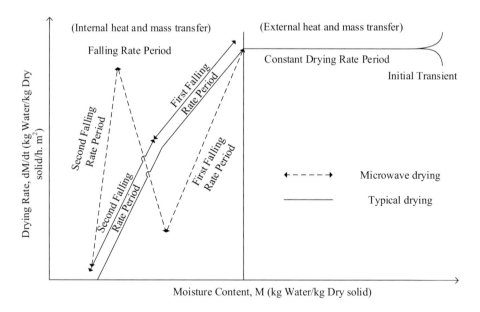

FIGURE 2.1   Drying characteristics of porous media.

TABLE 2.1

Classification of Drying Technology System

Classification	Descriptions	References
Drying strategy	Continuous, intermittent (on/off), cyclic (hot-cold/cold-hot), step wise (step up/step down)	Law et al. (2014) Chong et al. (2014)
Drying medium	Hot air, cold air, superheated steam, dehumidified air, inert gas, exhaust gas, osmotic solution	Law et al. (2014)
Handling of material	Stationary, packed, rotary, conveyor, fluidized bed, spouted bed, pulsated bed, vibrated bed, spray, freeze	Law et al. (2014)
Mode of input	Convection, conduction, vacuum, radiation (Solar/Vacuum), microwave, infrared, radio frequency, high electric field, pulsed electric field, high pressure, centrifugal force and ultrasound	Law et al. (2014) Ahmed et al. (2016)

wet, the drying process proceeds to the falling rate period, where moisture is diffused from the internal to the surface of material. The falling rate period itself is divided into two, starting from the initial or first falling rate period where the porous solid's surface is partially wet to when it becomes completely dry during the second falling rate period. However, the drying characteristics at the falling rate period can be modified by using different mode of heat input (for instance microwave drying) (Figure 2.1) or different mode of drying operation (for instances, intermittent drying or a variable drying parameter such as varying drying temperature etc.). During this period of the drying process determining the diffusion coefficient is possible.

There are different types of drying technology to prolong the shelf life of a product. Drying technology can be differentiated into many classifications, starting from drying strategy, drying medium, and handling of material to different mode of heat input (Table 2.1). There are more than 200 types of dryers and the drying parameters are what make them different from one another. Some authors tend to differentiate the use of "drying" for water removal process by exposure to sun and "dehydration" as artificial drying under controlled conditions. Among the technologies, osmotic dehydration, vacuum drying, freeze drying, superheated steam drying, heat pump dehumidifier drying, microwave drying and spray drying offer great scope of product quality. Several types of drying technology that are popular for drying porous media are discussed in this chapter.

2.1.1.1 Natural Sun Drying and Solar Drying

Natural sun drying and solar drying are among the oldest drying technologies and utilize solar energy to preserve agricultural surpluses. Natural sun drying and solar drying are mainly used for drying fruits and agricultural products like grapes, chempedak, ciku, salak, paddy, corn, fishes, prawns, crabs, cereals, apple, orange, mango, banana, pear, pineapple, guava, kiwi, dragon fruit, cocoa, dates and figs. Nowadays, solar drying systems have improved significantly compared to the last decade. Wang et al. (2018) used an indirect forced convection solar dryer (IFCSD) to dry mango and the effective diffusivity value ranged from 6.41×10^{-11} to 1.18×10^{-10} m²/s over the temperature range of $40 - 52°C$. Further to this, roof-integrated greenhouse solar and mixed-mode drying systems were introduced to reduce the drying time and to ensure a better quality of the product. The mixed-mode natural convection solar dryer is better than the natural convection solar dryers in drying of paddy. However, the natural convection is suitable for low-moisture-content crops.

2.1.1.2   Hot-Air Drying

Hot-air drying is the common drying method used in the industry. In general, air is used as the drying medium, heated to a certain temperature and brought in contact with the porous media to remove surface moisture and internal moisture. Agnihotri et al. (2017) reported that low relative humidity was found as a major controlling factor for drying rate except in high humidity conditions when the drying temperature falls between $30 - 60°C$. The effective moisture diffusivity for the *Inula racemose* rhizomes dried at $30 - 60°C$ and relative humidity at $30 - 80\%$ was reported to be ranged from 2.05×10^{-10} to 9.11×10^{-10} m^2/s.

Recent hot-air drying technology focuses on improving the product quality of samples dried using low-temperature dehumidified air and swirl flow. For example, the usage of a heat pump improves the thermal economy and efficiency of a conventional hot-air dryer. Malaikritsanachalee et al. (2018) proposed another way to improve the quality of dried product by using different type of airflow in hot-air drying of pineapple. They reported that the pineapple dried using hot-air drying was faster in swirling flow compared to non-swirling flow, and the effective diffusivity values ranged from 6.7×10^{-9} to 10.2×10^{-9} m^2/s. The highest drying rate can be obtained by setting the swirling flow type dryer at $70°C$ at air velocity of 2.0 m/s.

2.1.1.3   Osmotic Dehydration

Osmotic dehydration is used to partially remove water from a porous media using a hypertonic solution. Normally, osmotic agents used in osmotic dehydration are sucrose, corn syrup, salt, fructo-ologosaccharide, maltose, honey, maltodextrin, ethanol, high fructose corn syrup and maple syrup (Ahmed et al. 2016). The main objectives of using osmotic dehydration are to improve nutritional and functional value and organoleptic properties. This is because it can minimize the heat damage to the porous media and enhance drying rate. However, sugar gain of porous media during the osmotic dehydration is perceived negatively from a pro-health point of view.

Osmotic agents affected the diffusivity and drying rate of porous media. Some agents can reduce the change of porous media structure. For instance, salt provides the driving force for mass transfer and hinders the shrinkage of surface. When the surface volume changes are minimized, the change of porous structure is also minimized. To increase the diffusivity and drying rate during osmotic dehydration, maltose, high fructose corn syrup and honey can be used. For example, a honey solution has high osmotic pressure and high fructose corn syrup has a higher viscosity compare to other agents (Zhou and Jiang 2009; Sunjka and Raghavan 2004). These characteristics increased the water diffusion from porous media. However, fructo-oligosaccharide is another osmotic agent that lowers the diffusivity rate compared to sucrose due to its molecular weight (Matusek et al. 2008).

2.1.1.4 Freeze Drying

Freeze drying is used to produce high-value medicinal products and high-quality products. This technology involves freezing the porous media at low temperature instead of using high temperature drying medium as driving force. It can minimize overall structure damage and increase porosity that gives good rehydration characteristics. Highly porous materials such as those produced from freeze drying should be stored at lower relative humidity to prevent shrinkage after drying as shrinkage degree increased with the increase in relative humidity.

Freeze drying is good for producing high-porous-media samples but it is costly due to long freezing and drying duration. The selected modes of heat input stated in Table 2.1 can be used to increase the drying rate during freezing or to decrease drying time including radiation (solar/vacuum), microwave, infrared, radio frequency, high electric field, pulsed electric field, high pressure, centrifugal force and ultrasound. Combining other modes of heat input either intermittently or continuously can reduce the drying duration up to 60% compared to conventional freeze drying (Lombraña et al. 2001). However, this kind of heat input, especially microwave energy, needs accurate monitoring in order to avoid excessive heating and melting of the samples. In addition, poor mass transfer rate can lead to melting of samples.

2.1.1.5 Hybrid Drying/Combined Drying

Microwave drying is used to reduce drying duration at the falling rate period where the drying rate is normally very slow. Conventional drying methods cause sample surfaces to harden during the drying process and the effective diffusivity value decreased significantly at later stages of the falling rate period due to severe shrinkage. Microwave drying can create distinctive characteristics at the falling rate period region (Figure 2.1). The microwave drying utilizes radio frequency energy to generate heat inside the porous media. It can achieve a uniform heating of material because of radiation heat transfer and results in a better quality of final product. The effective diffusivity is increased with increases in microwave output powers and decreased with increases in samples volume. However, microwave drying causes the porous media surface to burn and shrink significantly when not properly used. It is advisable to combine microwave drying with vacuum so that the issues mentioned above can be overcome.

Hybrid drying techniques tend to offer better-quality dried products. They can be a combination of dryers or drying technologies. For instance, combining hot-air drying and freeze drying could increase the drying rate and offer advantages in producing better-quality dehydrated fruits and vegetables. Intermittent hot-air– and cold-air–drying techniques can be applied to dry heat sensitive products. For example, cold air can be supplied to the samples in the middle or at the beginning of the drying process or osmotic dehydration can be used as a pretreatment. Recently, Dehghannya et al. (2018) dried quince fruit using multi-stage continuous and intermittent microwave drying with osmotic dehydration and low-temperature hot-air drying. During the drying process, osmotic pre-treated quince fruit were dried using intermittent microwave (IM) combined with hot-air (HA) drying at 40°C. The osmotic solutions were prepared at five different concentration levels of from 0 (control) to 70% (w/w) and four different microwave powers ranged from 0 (control) to 900 W, with four pulse ratios of 1, 2, 3 and 4. The authors reported that the effective diffusivity values increased by increasing the microwave power and pulse ratio, while they decreased by increasing the concentration of the osmotic solution. The effective diffusivity value obtained from falling rate period varied from 3.23×10^{-10} to 7.82×10^{-10} m²/s.

Microwave drying technology is also used as an alternative assisted drying technology with hot air. Combining hot air and vacuum-microwave (HA/VM) drying technology produced porous media at lower final moisture content and higher consumer preference. The hybrid drying exhibits an unusual drying curve where it starts from a constant rate period, followed by the first falling rate period, and then the second falling rate period. A great increment after the first falling rate periods occurs due to the change of mode of input. It is caused by the rapid water removal from the internal parts of porous media due to the microwave energy.

Another example is microwave hot-air drying. Sharma and Prasad (2004) dried garlic cloves at 10 – 40 W at intervals of 10 W. At the same time, the hot air was set at 40, 50, 60 and 70°C at velocities of 1.0 and 2.0 m/s. The average effective diffusivity increased from 1.6×10^{-10} to 9.7×10^{-10} m²/s when the air temperature increased from 40 to 70°C (air velocity = 1.0 m/s).

2.1.2 DRYING OF POROUS MEDIA

In this chapter, five porous materials (fruits, foods, wood, coal and ceramic) are discussed in detail. There are many ways to classify a porous material, from the size or number of pores to the structure of the pores itself. Though there are many types of porous material, they share some common characteristics, which are low relative density, large specific surface area, high specific strength and small thermal conductivity.

A porous medium is described as a solid containing many holes or pores and tortuous passages. The characteristics of a porous medium are determined by the manner in which the pores are embedded for example the blind pores, interconnected pores, closed pores and pores with different size. The fraction of the medium that contains voids is known as its porosity, and pores are formed from a fluid phase. There are many materials that can be classified as porous media, but not all materials with pores can be referred as porous. Porous materials ought to possess two essential characteristics, which are that the material contains many pores and that the pores are formed specifically to meet a certain size that the material can be used. Figure 2.2 show the characteristics of porous structures.

2.1.2.1 Drying Porous Media (Fruits)

Fruits are mainly composed of crumb tissue called parenchyma. The formations of the crumb tissue are usually quite large, with a diameter range from 0.01 to 0.5 mm. The characteristic feature of parenchyma is the intercellular spaces are filled with water and air; the intercellular spaces provide all the cells with oxygen and allow gas exchange with the atmosphere. The distribution of the intercellular space is anisotropic and the properties depend on the type of fruit, its species and variety. For example, the shapes of the cells in apple tissue are highly dependent on their function and location.

Processing fruits by drying causes significant change to the inner structure of fruits; the heat and mass exchange process can cause changes in cell walls that lead to

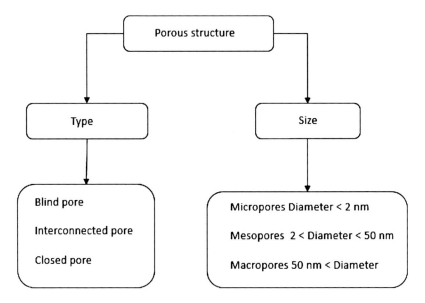

FIGURE 2.2   Characteristics of porous structures.

shrinkage and change the fruit's structure. The degree of the structural changes of the pore depends on the distribution of the water, which is governed by the mechanism of internal moisture transport. It starts with the movement of intercellular water followed by the intracellular water movement leading to cell shrinkage and pore shrinkage. Further, the drying process causes cell collapse and enlargement of pores. Various type of drying technologies have been used to preserve fruits, the effective diffusivity values of different fruits dried using different drying technologies from conventional drying to combined/advanced drying typically ranged from 10^{-10} to 10^{-8} m²/s.

2.1.2.2   Drying of Porous Media (Coal)

Coal is an example of porous media. Coal meso or macropore structure changes during drying. The porous structure of low-rank coals is comprised of meso- and macropores and low-rank coals are dominated by mesoporous structure. Yu et al. (2013) reported that slit-like pores of moistened coal contain bound water and non-freezable water. After the drying process, the change of structure affected the flow of gaseous reactants and products such as oxygen, carbon dioxide, hydrogen, carbon monoxide and methane. Several drying methods have been used to dry low-rank coals including convective air drying, superheated steam drying and microwave drying. There is an impact from the drying temperature perspective; some researchers have reported that the absorption of carboxyl (COOH) and carbonyl (C=O) groups decreased with increasing drying temperature. This is due to the progressive loss of oxygen functional groups with increasing drying intensity and microwave output level during superheated steam drying and microwave drying, respectively (Yu et al. 2013).

Low-rank coals exhibited constant period and falling rate period drying characteristics. The pore shrinkage and emptying occurred due to the counteractions of particle contraction and moisture removal. Evans (1973) reported that Yallourn brown coal contains 200 g of water per 100 g of dry coal. Water molecules present in large pores were first removed by evaporation followed by those in the large capillaries. Shrinkage only occurred when the moisture content dropped to 40 g per 100 g dry coal due to the open gel structure collapse. Kelemen et al. (2006), found that drying at 150°C significantly alters the connectivity of pore structure network of coal. Some researchers proposed using emerging drying technology. Superheated steam drying, microwave drying or combined drying methods have been proposed to improve the drying of coal.

Microwaves can generate high energy densities and temperatures in the coal. According to Pickles and Kelebek (2014), the average effective diffusion values ranged from 6.45×10^{-7} m²/s at a mass-to-power ratio of 0.1875 W/g to 2.84×10^{-6} m²/s at a mass-to-power ratio of 0.0536 W/g. Generally, the lignite coal effective diffusivity value ranged from 2×10^{-8} to 8×10^{-8} m²/s for moisture fractions ranging from 0 to 1 and temperatures from 25 to 200°C (Pakowski et al. 2012). In terms of drying duration, it was found that the drying duration for convective air drying was about 3000 s at moisture fraction of 0.31. However, a sample dried using microwave drying completed in 300 s at moisture fraction of 0.014 (Pickles and Kelebek 2014).

2.1.2.3   Drying of Porous Media (Foods and Vegetables)

Of the wide range of foods that can be categorized as porous media, one of the examples is the cereal cellular products. The cereal cellular products consist of a solid matrix of mainly starch with interconnected air cells dispersed in it. Crispness is the essential quality attribute of this kind of food product and this textural attribute tends to be lost when they absorb too much moisture.

Moisture transfer depends highly on the structure of the material. Esveld et al. (2012) studied the difference in relative vapor conductivity of fine- and coarse-structured crackers. The authors found that the difference in porosity is not due to their difference in morphology scale, but to the difference in open surface fraction between cells.

The effective diffusivity of foods depends on different drying technologies. Kraus et al. (2013) reported that the effective diffusivity values of starch-based pellets dried using microwave drying ranged from $3.4 \times 10^{-9} - 1.2 \times 10^{-8}$ m²/s. In drying parboiled wheat, Kahyaoglu et al. (2012) used microwave-assisted spouted-bed drying at 3.5 W/g and 7.5 W/g. This drying technology can reduce the drying time by at least 60% and 85% compared to conventional drying technology. In term of effective diffusivity, it ranged from $1.44 \times 10^{-10} - 3.32 \times 10^{-10}$ m²/s.

Other types of foods that exhibit the porous structure are vegetables. Drying technology preserves and improves the stability of the vegetables, but the process affected their nutritional values. The drying rate of vegetables is governed by the internal moisture diffusion, internal structure and other drying parameters. The main challenge is shrinkage of vegetables type, which represent porous media during drying. The degree of shrinkage is different depending on the type of vegetables and type of drying as well as the drying condition. The shrinkage values affected the effective diffusivity values. Factors like temperature, moisture content, product thickness and vacuum pressure also affected the drying rate of vegetables.

Combined drying technology can increase the diffusivity values of vegetable-type porous media. Aydogdu et al. (2015) claimed that hot-air dried eggplants had less porous structure than ones dried in microwave and infrared combination oven. The effective diffusivity of the eggplant was 5.1×10^{-10} m²/s for hot-air drying and around $7.1 \times 10^{-9} - 14.45 \times 10^{-9}$ m²/s for microwave-infrared combined drying. In addition, drying duration was reduced significantly using the combination of infrared and hot-air drying technology.

2.1.2.4   Drying of Porous Media (Woods)

Timber is defined as wood harvested from trees that may be converted and used to manufacture wood composites. Wood is considered a porous media with complex structure due to different anatomical structures. In general, there are two categories of commercial timber, which are the softwoods and hardwoods. Softwoods are made up only of cells like tracheids and rays, while hardwoods are arranged by more types of cell in more varieties of arrangement. In evolutionary terms, hardwoods are younger and consist of more complex structures compared to softwoods. The weight of a piece of wood is not constant; wood can both swell and shrink as it gains and loses moisture. The moisture in wood can be either internal moisture in the wood cell wall or external water in cell lumen. The term "fiber saturation point" is defined as the moisture content at which cell lumens are free of water, but the cell walls are still fully saturated. The fiber saturation point ranges from 20 to 25% depending on the species of tree. Moisture inside the wood need to be removed to increase the quality of the wood, hence a drying process is applied for this purpose. When wood dries, water evaporates from the lumens and intercellular spaces and then water molecules are absorbed by the dry cell walls leading to the swelling process. Figure 2.3 shows a cell wall and cell lumen of a wood.

Drying woods is a complicated process; the physical and mechanical behavior of the wood varies during the drying process. Defects to the inner structure that caused collapsing can occur due to high drying stress. Other parameters like moisture

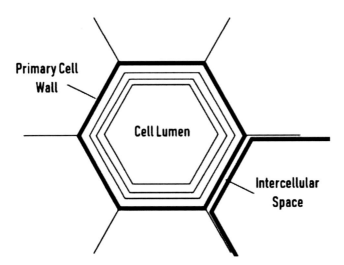

FIGURE 2.3   Cell wall and cell lumen.

content, temperature and the type of drying also affect the process of drying; studies also reported that wood properties influence the movement of moisture during dying.

Convective drying is commonly used to dry woods. Sridhar and Madhu (2015) claimed that the drying rate increased with drying temperature in the convective drying of *Casuarina Equisetifolia* wood chips, and the effective diffusion coefficient increased from 4.43×10^{-9} m²/s to 1.01×10^{-8} m²/s when the temperature increased from 80 to 100°C.

Another type of drying that is also commonly used for drying woods is vacuum drying. Vacuum drying offers various advantages like high drying rate at lower temperature (lower boiling point), shorter drying times, color preservation and better control of volatile organic compound emissions. However, it is usually combined with other drying technology or pretreatment technology. Steam, ultrasound and microwave mode of inputs are applied in vacuum drying of wood. The microwave conditioning produces better product quality compared to steam conditioning (He and Wang 2015). This is because during the microwave heating process, the heat absorbed drives internal water to the surface of the wood structure. He et al. (2017) also reported that microwave pretreatment enhanced permeability by effectively decreasing the moisture content; drying rate also accelerated by 171% compared to the untreated sample. Other methods like ultrasound-assisted vacuum drying show a faster drying and higher effective water diffusivity than the normal hot-air drying when removing free water.

2.1.2.5   Drying of Porous Media (Ceramic)

Ceramics are used for a wide variety of applications from filtration, separation, catalysts, and catalysts supports to lightweight structural components. It has low density, high surface area and good thermal shock resistance; high specific strength properties can be used for high temperature and corrosive environment. Many researchers in this field are focusing on tailoring sizes, shapes, amounts and connectivity of distributed pores to control the porous structure and improve their function. The process of drying ceramics is generally longer compared to other porous media to maintain the product quality. The internal moisture movement controls the drying rate and the drying system needs to be maintained at a certain temperature to prevent deformation and cracking. Itaya et al. (2007) compared the quality of ceramics dried using microwave, hot air and radiation as heat inputs. The authors reported that the microwave drying could reduce drying duration but the microwave power needs to be controlled so the temperature is less than the boiling point of water to avoid cracking the ceramics. In terms of drying characteristics, Capela et al. (2016) claimed that high-porosity ceramic composites exhibited the first falling rate period and the second falling rate period in the entire drying process. In drying of ceramic, the volumetric shrinkage is less than 5%. Unidirectional drying is carried out to allow homogenous shrinkage in the foam and avoid formation of pores inside the body. Drying technology is also applied to fabricating a ceramic membrane.

For ceramic membrane preparation, drying is an important factor associated with the failure of the consolidated structure. Therefore, a more advanced drying technology is used to fabricate high-quality ceramic. Wang et al. (2017) fabricated porous Si_3N_4 ceramics using incomplete gel casting and freeze-drying technology. The authors claimed that side-surface heating boundaries drying rates are relatively low due to the generation of higher pore water pressure and gas pressure in a solid matrix. In addition, the drying of hygroscopic layer can result in higher moisture gap and lead to increase cracking possibilities.

2.2 DIFFUSION DURING DRYING OF POROUS MEDIA

The movement of molecules in a fluid due to the concentration gradient is known as diffusion, and the degree of moisture movement of every material is known as its diffusion coefficient. Molecular diffusion occurs in various mass transfer processes, for example the drying of porous media. The molecular diffusion that occurred in drying of porous media mainly are moisture movement in liquid form. There are various types of moisture diffusion that can occur inside the structure of porous media during the drying process. Porous media have been used in many aspects of our lives, ranging from foods to building materials like wood, coal, ceramics and more. The complex structure of porous material is characterized using the term porosity. The complex structure affects the drying kinetics and characteristics of porous media. Figure 2.4 shows the entity relationship diagram (ERD) between drying technology and porous media. Physical state, structure and chemical composition of the porous media change along the drying process, for example exhibiting shrinkage, puffing, etc. The mechanisms of heat and mass transfer are affected by the changes of physical state and other properties. Considering the fact that even the same porous material will not have an identical internal structure, the data obtained from other resources may not be used directly. In addition, parameters like temperature, initial moisture content and others also affect the drying rate.

2.2.1 FICK'S FIRST AND SECOND LAW

Diffusion is defined as the random movement of molecules in a fluid from a high concentration region to a lower concentration region; the driving force of diffusion is

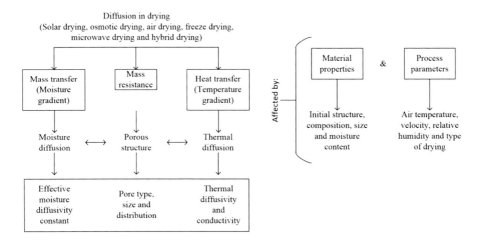

FIGURE 2.4 Entity relationship diagram (ERD) of drying technology and porous media.

the concentration gradient. Fick's law of diffusion can characterize diffusion. Fick's first law relates the flux of molecules to the concentration gradient (Eq. 2.1):

$$J = -D\frac{\partial c}{\partial x}$$ (2.1)

Where the diffusive flux J is the amount of substance per unit in time (mol/m².s); D is the diffusion coefficient (m²/s), also referred as the ability of a substance to diffuse; c is the concentration (mol/m³) and x is the position (m). The negative sign shows that the flux is in the direction of decreasing concentration and is driven in the direction of increasing of the position.

Fick's second law describes the change of concentration over the time (Eq. 2.2):

$$\frac{\partial c}{\partial t} = D\frac{\partial^2 c}{\partial x^2}$$ (2.2)

Where c is the concentration (mol/m³), D is the diffusion coefficient (m²/s), t is the time (s), and x is the position (m).

Drying is a simultaneous heat and mass transfer process. That is, there are two types of diffusion in the drying process, which are moisture diffusion and thermal diffusion (Figure 2.4).

2.2.2 MOISTURE AND THERMAL DIFFUSION

Diffusion is defined as the movement of atoms, molecules or clusters of molecules in vapor, gases, liquid and solids. Moisture may present as external moisture at the surface or internal moisture within the material. In terms of fluid flow, moisture diffusion of drying can be categorized into external and internal moisture diffusion. External moisture diffusion is where the water molecules at the surface of the material are transferred to the surroundings and the internal moisture diffusion is where the water molecules within the porous material are transferred to the surface. The movement of liquid water and water vapor may occur by any one or more of the following mechanisms: liquid diffusion, vapor diffusion, Knudsen diffusion and capillary flow. Internal liquid water can migrate to the surface through liquid diffusion and/or capillary flow. A moisture gradient and/or temperature gradient between the surface and interior is the driving force of liquid diffusion, while the molecular attraction between surface and water molecules causes capillary suction. The movement of internal water vapor in the drying process may occur through vapor diffusion and/or Knudsen diffusion. As the heat energy is transferred to the material, up to a point where the boiling point of the liquid water is met, water vapor is formed and then diffuses to the surface. Knudsen diffusion only occurred at certain condition where the temperature and pressure is very low, for example in freeze drying. The internal water vapor flow through the Knudsen diffusion mechanisms occurs when the mean free path of the vapor molecules is greater than or equal to the diameter of the pore. At the surface, the liquid water and water vapor migrates away from the surface through liquid diffusion and vapor diffusion. The effective moisture diffusivity constant is the term used to characterize the ability of moisture to move in the material. The effective moisture diffusivity constant of every porous material is different due to the complex structure of porous material. It is affected by the internal pore structure, moisture gradient, type of drying and drying parameters like temperature, relative humidity and more (Figure 2.4).

The transport of heat in the drying process occurs primarily by conduction and convection, though radiation heat transfer may occur for certain types of drying such as infrared or solar drying. Heat transfer takes place due to the temperature gradient. Heat transfer through conduction occurs through the contact of two different solid material or two regions of the same material with different temperature (Eq. 2.3).

$$Q = -K A \left(\frac{\partial T}{\partial x} \right) \qquad (2.3)$$

Where Q is the heat flux (J/s = W); K is the thermal conductivity (W/m.K), also known as the ability of a material to conduct heat; A is the surface area (m^2) and $\frac{\partial T}{\partial x}$ is the temperature gradient (K/m).

Heat transfer through convection occurs through movement of fluids (Eq. 2.4).

$$Q = h A \Delta T \qquad (2.4)$$

Where Q is the heat flux (J/s = W), A is the surface area (m^2), h is the heat transfer coefficient or film coefficient (W/m^2·K), which depend on the film of the fluid, and ΔT is the temperature difference (K).

Another thermophysical property known as thermal diffusivity is used to describe the extent of how a material diffuses or spreads heat throughout its body (Eq. 2.5).

$$\propto = \frac{K}{Cp \ x \ \rho} \qquad (2.5)$$

Where $\propto$ is the thermal diffusivity (m^2/s), K is the thermal conductivity (W/m·K), C_p is the specific heat capacity (J/kg·K) and ρ is the density (kg/m^3). The typical thermal diffusivity ($\propto$) and thermal conductivity (K) values of fruits, cereal, coal and ceramic range from 1.0×10^{-7} to 3.0×10^{-7} m^2/s and 0.1 to 0.7 W/m·K, respectively.

2.3 DIFFUSION MODELS IN DRYING OF POROUS MEDIA

The mathematical modelling of diffusion in drying porous media is a useful tool in optimizing the drying process and in designing a better drying system. The complexity of porous media structure, chemical composition, transport phenomena and biological variability are taken into account in the drying process of porous media. Castro et al. (2018) described a number of parameters that must be defined for the mathematical modelling of drying of fruit (Table 2.2). The authors divided the parameters into transport phenomena, scale, thermophysical properties, transfer coefficient, solution methods and other considerations. The simplest models used to determine the diffusivity of porous media are thin-layer models. They are often used to describe biomaterial and are categorized into theoretical, semi-theoretical and empirical models.

TABLE 2.2

Parameters of Mathematical Modelling of Fruit Drying in 1, 2 or 3 Dimensions and Geometry

	Parameters
Transport phenomena	Mass, heat and momentum transfer with and without conjugate
Scale	Macro, micro and multi
Thermophysical properties	Isotropic, anisotropic and variable properties
Transfer coefficients	Empirical models, semi-empirical correlations, optimization and computational fluid dynamics (CFD) simulation
Solution methods	Analytical, numerical: finite difference, finite volume, finite elements etc.
Other considerations	Shrinkage, texture and other quality changes

Info adapted from Figure 1 of the journal paper. Copyright License number: 4416850096065; Date: Aug 26, 2018, Elsevier

An empirical model is based on fitting experimental data and performing dimensional analysis, and thus omits the fundamental drying process. The coefficients of the model have no physical meaning and they are determined based on coefficient of determination, chi square, mean square errors and root mean square error. The model with the highest coefficient of determination and lowest chi square, mean square errors and root mean square error is selected as the best empirical model to describe the drying kinetics.

Semi-theoretical models are derived from the theoretical models or their simplified variations. Semi-theoretical models can be classified according to their derivation, such as Newton's law of cooling and Fick's second law of diffusion. The models are the Lewis model, the Page model and modified forms, the single-term exponential model and modified forms, the two-term exponential model and modified forms and the three-term exponential model (Table 2.3).

Theoretical models gives a better understanding of the drying behavior compared to semi-theoretical and empirical models, but the limitation of the theoretical model is to only consider the internal resistances in the transfer of moisture, thus it is inadequate and

TABLE 2.3

Sample Models

No	Models	Mathematical equations
1	Newton Model	$MR = exp(-kt)$
2	Page Model	$MR = exp(-kt^n)$
3	Modified Page (II)	$MR = exp[-(kt)^n]$
4	Modified Page (III)	$MR = k\ exp\ (-t/d^2)^n$
5	Midilli and others model	$MR = a\ exp(-kt) + bt$
6	Two-term model	$MR = a\ exp(-k_1t) + b\ exp(-k_2t)$
7	Logarithmic model	$MR = a\ exp(-kt) + c$

tends to generate erroneous results due to the assumptions made. Fick's second law of diffusion is the most commonly used theoretical model in modelling the drying process.

Thin layer refers to a layer of small product thickness, where air characteristics are identically uniform everywhere in the layer. Thin-layer mathematic models are used to predict drying kinetics of porous media. Typical models are the Newton model, Page model, modified Page model, Midilli model, two-term model and so on. Recently, in a review, Onwude et al. (2016) compared 22 thin-layer models used to fit the drying kinetics of fruits and vegetables using conventional convective air drying technology. He reported that semi-theoretical models of Newton's law of cooling and Fick's second law of diffusion such as Midilli and others, Page, two-term, logarithmic, modified Page and the approximation of diffusion models are best-fit models in describing the drying behavior of various fruits and vegetables respectively. Resende et al. (2018) reported that the logarithmic model was the best fit to describe the drying process of baru fruits (*Dipteryx alata* Vogel). The effective diffusivity values ranged from 1.2×10^{-10} to 11.9×10^{-10} m²/s at elevated temperature (40 – 100°C). In terms of wood, empirical models such as the logarithmic and modified Henderson and Pabis models predicted the drying rate in the best manner among the models used in the temperature range 80 – 100°C, with the effective moisture diffusivity ranged from 4.43×10^{-9} m²/s – 10.15×10^{-9} m²/s. Theoretically, the effective diffusivity values for wood ranged from 8.72×10^{-10} m²/s to 1.1×10^{-9} m²/s (Gebgeegziabher et al. 2013).

For hybrid drying technology, Sampaio et al. (2017) found that Midilli's model best fit the drying kinetics of persimmon using the osmo-convective drying. The experimental data were conducted at temperature of 50, 55 and 60°C and the effective moisture diffusivity was reported as 1.04×10^{-6}, 1.4×10^{-6}, and 8.27×10^{-7} m²/s, respectively.

Vu and Tsotsas (2018) reviewed the mass and heat transport models for analysis of drying process in porous media. Six theoretical models were assessed and compared, starting from the simplest diffusion model to the most complicated Whitaker's model (Figure 2.5).

Referring to Figure 2.5, the diffusion model is the simplest model among the theoretical models, due to the heat and mass transport only being defined from a single mechanism, which is the diffusion of vapor. In reality, the moisture movement is due to not only diffusion but also to other mechanisms like capillary suction, convection, external pressure, etc. Whitaker's model is proven to describe the drying process the best as it incorporates heat and mass transfer mechanisms. However, it is challenging when it comes to determining the transport coefficients of the Whitaker's model.

The complexity of the model is due to the inclusion of the mechanisms of the heat and mass-transport components. Each models are categorized at certain level of difficulty from heat and mass transfer perspective during the drying process.

FIGURE 2.5 Theoretical models.

Diffusion model equation:

$$\frac{\partial M}{\partial t} = \nabla.(D_{eff}\nabla M) \tag{2.6}$$

Where M is defined as moisture content, D_{eff} is effective diffusivity value and t is time.
Whitaker's model
Water in both liquid and gas phases

$$\frac{\partial}{\partial t}\left(\rho_w \varepsilon_w + \varepsilon_g \rho_v\right) + \nabla.\left(-\rho_w.\frac{Kk_w}{\eta_w}\nabla P_w - \rho_v \frac{Kk_g}{\eta_g}\nabla P_g\right) = \nabla.\left[\rho_g D_{eff}.\nabla\left(\frac{\rho_v}{\rho_g}\right)\right] \tag{2.7}$$

Conservation equation for air in the gas phase

$$\frac{\partial}{\partial t}\left(\varepsilon_g \rho_a\right) + \nabla.\left(-\rho_a \frac{Kk_g}{\eta_g}\nabla P_g\right) = \nabla.\left[\rho_g D_{eff}.\nabla\left(\frac{\rho_a}{\rho_g}\right)\right] \tag{2.8}$$

Conservation equation of energy

$$\frac{\partial}{\partial t}\left(\varepsilon_s \rho_s h_s + \varepsilon_w \rho_w h_w + \varepsilon_g \rho_v h_v + \varepsilon_g \rho_a h_a\right) +$$

$$\nabla.\left[-\rho_w.\frac{Kk_w}{\eta_w}\nabla P_w.h_w - \left(\rho_v h_v + \rho_a h_a\right).\frac{Kk_g}{\eta_g}\nabla P_g\right] = \tag{2.9}$$

$$\nabla.\left[\rho_g h_a D_{eff}.\nabla\left(\frac{\rho_a}{\rho_g}\right)\right] + \nabla.\left[\rho_g h_v D_{eff}\nabla\left(\frac{\rho_v}{\rho_g}\right)\right] + \nabla.\left(\lambda_{eff}\nabla T\right)$$

Where the ρ_w, ρ_v and ρ_g are the mass density of the liquid, vapor and gas phases; ε_w and ε_g are the volume fractions of the liquid and gas phases; η_w and η_g are the dynamic viscosity of water and gas; k_w and k_g are the relative permeability tensors of liquid and gas, K is the absolute permeability tensor and D_{eff} is the effective diffusivity tensor.

2.4   CONCLUSIONS

In drying porous media, drying kinetics study and modelling are important to understand the drying characteristics and drying process as well as when designing a dryer. Combining different drying strategies, drying mediums and mode of input create multiple distinctive periods in the drying rate curve (drying-rate–versus–moisture-content plot). The period can be falling rate periods or increasing rate periods, depending on the drying strategy, drying parameters and mode of heat input used in the drying process. Obvious inflection point can be clearly seen from the drying rate curve of porous media at the end of the process. The greatest increment in drying rate could be found in porous media dried using hybrid drying techniques assisted by microwave drying towards the end of the drying process. This is due to the rapid removal of water from the internal structure of porous media by microwave energy. Combined/advanced drying technologies tend to reduce drying duration and optimize the process and thus are more suitable for application in industry.

In terms of modelling, the coefficients of empirical and semi-empirical models can be used to distinguish critical moisture contents at different periods. The critical moisture content is important when we wish to apply a two-stage drying strategy. The moisture content would mark the onset of the second drying technique. The model selected is highly dependent on the empirical data obtained from experiments or scale-up data. As long as the drying medium and mode of input are similar, the models can predict the moisture content very well. For the theoretical model, Whitaker's model can be considered as the best as it includes basic law of mass and heat transport at the macroscopic level. It incorporates all mechanisms, therefore it is more complicated but simultaneously it is more accurate.

REFERENCES

Agnihotri, V., Jantwal, A. and Joshi, R. 2017. Determination of effective moisture diffusivity, energy consumption and active ingredient concentration variation in *Inula racemosa* rhizomes during drying. *Industrial Crops & Products* 106:40–47.

Aydogdu, A., Sumnu, G., and Sahin, S. 2015. Effects of microwave-infrared combination drying on quality of eggplants. *Food and Bioprocess Technology* 8:1198–1210.

Ahmed, I., Qazi, I. M. and Jamal. S. 2016. Developments in osmotic dehydration technique for the preservation of fruits and vegetables. *Innovative Food Science and Emerging Technologies* 34(C):29–43.

Capela, P., Faria, L., Junior, C., Carvalho, L., Guedes, A., Pereira, M., and Soares, D. 2016. Study and optimization of the drying process of a ceramic abrasive composite. *International Journal of Applied Ceramic Technology* 13:308–315.

Castro, A. M., Mayorga, E. Y., and Moreno. F. L. 2018. Mathematical modelling of convective drying of fruits: A review. *Journal of Food Engineering* 223:152–167.

Chong, C. H., Figiel, A., Law, C. L. and Wojdyło, A. 2014. Combined drying of apple cubes by using of heat pump, vacuum-microwave, and intermittent techniques. *Food and Bioprocess Technology* 7, 975–989.

Dehgahannya, J., Hosseinlar, S. and Heshmati, M. K. 2018. Multi-stage continuous and intermittent microwave drying of quince fruit coupled with osmotic dehydration and low temperature hot-air drying. *Innovative Food Science and Emerging Technologies* 45:132–151.

Esveld, D. C., van der Sman, R. G. M., van Dalen, G., van Duynhoven, D. P. N., and Meinders, M. B. J. 2012. Effect of morphology on water sorption in cellular solid foods. Part I: Pore scale network model. *Journal of Food Engineering* 109:301–310.

Evans, D.G. 1973. The brown-coal/water system: Part 4. Shrinkage on drying, *Fuel* 52:186–190.

Gebgeegziabher, T., Oyedun, A. O., Zhang, Y., and Hui, C.W. 2013. Effective optimization model for biomass drying. *Computer Aided Chemical Engineering* 32:97–102.

He, Q., and Wang, X. 2015. Drying stress relaxation of wood subjected to microwave radiation. *BioResources* 10:4441-4452.

He, X., Xiong, X., Xie, J., Li, Y., Wei, Y., Quan, P., Mou, Q., Li, X. 2017. Effect of microwave pretreatment on permeability and drying properties of wood. *Bioresources* 12:3850–3863.

Itaya, Y., Uchiyama, S., and Mori, S. 2007. Internal heating effect and enhancement of drying of ceramics by microwave heating with dynamic control. *Transport in Porous Media* 66:29–42.

Kahyaoglu, L. N., Sahin, S., & Sumnu, G. 2012. Spouted bed and microwave-assisted spouted bed drying of parboiled wheat. *Food and Bioproducts Processing* 90:301–308.

Kelemen, S. R., Kwiatek, L. M., Siskin, M., and Lee, A. G. K. 2006. Structural response of coal to drying and pentane sorption. *Energy & Fuels* 20:205–213.

Kraus, S., Sólyom, K., Schuchmann, H., and Gaukel, V. 2013. Drying kinetics and expansion of non-predried extruded starch-based pellets during microwave vacuum processing. *Journal of Food Process Engineering* 36:763–773.

Law, C. L., Chen, H. H. H., Mujumdar, A. S. 2014. Food Technologies: Drying. In *Encyclopedia of Food Safety*, ed. Y. Motarjemi, M., Gerald, and T. Ewen, 3:156–167. Elsevier, B.V.

Lombraña, J. I., Zuazo, I., Ikara, J. 2001. Moisture diffusivity behavior during freeze drying under microwave heating power application. *Drying Technology* 19:1613–1627.

Malaikritsanachalee, P., Choosri, W., Choosri, T. 2018. Study on kinetics of flow characteristics in hot-air drying of pineapple. *Food Science and Biotechnology* 27:1047–1055.

Matusek, A., Czukor, B., & Meresz, P. 2008. Comparison of sucrose and fructo oligosaccharides as osmotic agents in apple. *Innovative Food Science & Emerging Technologies*, 9:365–373.

Onwude, D., Hashim, N., Janius, R., Nawi, N., and Abdan, K. 2016. Modeling the thin-layer drying of fruits and vegetables: A review. *Comprehensive Reviews in Food Science and Food Safety* 15:599–618.

Pakowski, Z., Adamski, R., Kwapisz, S., 2012 Effective diffusivity of moisture in low-rank coal during superheated steam drying at atmospheric pressure. *Chemical and Process Engineering* 33:43–51.

Pickles, C. A., Kelebek, F. G. S. 2014. Microwave drying of a low-rank sub-bituminous coal. *Minerals Engineering* 62:31–42.

Resende, O. M., Costa, L. N., Ferreira Júnior, W. E., and De Oliveira, D. 2018. Drying kinetics of baru fruits (*Dipteryx alata* Vogel). *Engenharia Agricola* 38:103–109.

Sampaio, R. M., Neto, J. P. M., Perez, V. H., Marcos, S. K., Boizan, M. A., and Da Silva, L. R. 2017. Mathematical modeling of drying kinetics of persimmon fruits (*Diospyros kaki* cv. Fuyu. *Journal of Food Processing and Preservation* 41:1–7.

Sharma, G.P., and Prasad, S. 2004. Effective moisture diffusivity of garlic cloves undergoing microwave-convective drying. *Journal of Food Engineering* 65:609–617.

Sridhar, D., and Madhu, G. 2015. Drying kinetics and mathematical modeling of *Casuarina Equisetifolia* wood chips at various temperatures. *Periodica Polytechnica-Chemical Engineering* 59:288–295.

Sunjka, P. S., and Raghavan, G. S. V. 2004. Assessment of pretreatment methods and osmotic dehydration for cranberries. *Canadian Biosystems Engineering* 46: 45–48.

Vu, H. T., & Tsotsas, E. 2018. Mass and heat transport models for analysis of the drying process in porous media: A review and numerical implementation. *International Journal of Chemical Engineering* 2018:1–13. http://hindawi.com/journals/ijce/2018/9456418

Wang, F., Gu, H., Yin, J., Yao, D., Xia, Y., Zuo, K., Liang, H., Ning, C., and Zeng, Y. 2017. Porous Si_3N_4 ceramics fabricated through a modified incomplete gelcasting and freeze-drying method. *Ceramics International* 43:14678–14682.

Wang, W., Li, M., Hassanien, R. H. E., Wang, Y. and Yang, L. 2018. Thermal performance of indirect forced convection solar dryer and kinetics analysis of mango. *Applied Thermal Engineering* 134:310–321.

Yu, J., Tahmasebi, A., Han, Y., Yin, F., Li, X. 2013. A review on water in low-rank coals: The existence, interactions with coal structure and effects on coal utilisation. *Fuel Processing Technology* 106:9–20.

Zhou, W., and Jiang, X. 2009. Process for treating plant material. (WO2009105039). https://patentscope.wipo.int/search/en/detail.jsf?docId=WO2009105039&tab=PCTBIBLIO&office=&prevFilter=&sortOption=Pub+Date+Desc&queryString=FP%3A%28200910 5039%29&recNum=2&maxRec=3

3 Multiscale Modeling of Porous Media

Peng Xu
China Jiliang University, Hangzhou, Zhejiang, China

Arun S. Mujumdar
McGill University & Western University,
Montreal, Quebec, Canada

Agus P. Sasmito
McGill University, Montreal, Quebec, Canada

Boming Yu
Huazhong University of Science and
Technology, Wuhan, Hubei, China

CONTENTS

3.1   INTRODUCTION

Drying is a very important industrial operation accounting for ten to twenty-five percent of the total energy used in the manufacturing process worldwide (Mujumdar and Passos, 2000). Thereinto, drying of porous materials is involved in the production process of a variety of commercial products such as food, paper, textile, wood, ceramics, granular and building materials, pharmaceuticals etc. (Tsotsas and Mujumdar, 2007). However, the drying process in porous media is a rather complex

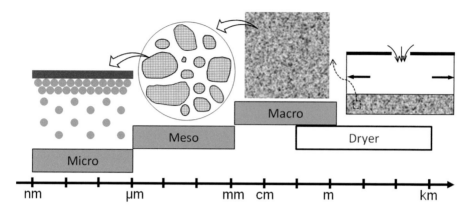

FIGURE 3.1   Multiscale nature of drying of porous media.

phenomenon as multiphase and multiphysics transport processes at multiple scales maybe involved (Kowalski, 2003; Defraeye, 2014).

Porous material is generally composed of extensive pores and solid matrix, and drying of porous media has a distinct *multiscale* nature. Typical scales involved in drying porous media are shown in Figure 3.1. It should be noted that these scales may overlap to some extent and the dividing method may be different depending on the specific porous material. The microstructures, including pore/particle shape and size distribution, arrangement of pores, pore space connections, tortuosity of the flow paths, composition of pores etc., are extremely complicated; it is of great importance to study the transport mechanisms and predict physical properties such as porosity, saturation, thermal conductivity, diffusion coefficient, permeability etc. of porous media. As a process of moisture removal from saturated materials, drying of porous media involves *multiphases*, including solid matrix, pore spaces, water, gas etc. The liquid water inside the pores can be classified into free water and irreducible water (bound water). Free water can flow through a pore and capillary network, while irreducible water may contain physically-adsorbed water (physisorption by weak van der Waals forces) and chemically-adsorbed water (chemisorption via covalent bonding). The gas phase in drying of porous media may involve dry air, water vapour and other gasses. Drying of porous media is a *multiphysics* process as various heat and mass transfer mechanisms may take place simultaneously inside of the porous material and at the interface between the porous material and the external environment. The multiphysics process is generally composed of a series, parallel, and/or series–parallel combination of heat conduction, heat convection, heat radiation, phase change (evaporation), laminar and turbulent flow, seepage flow, gas diffusion, absorption-desorption etc. The multiscale, multiphase and multiphysics nature of porous materials makes modeling the transport physics in drying of porous media one of the major challenges in drying technology. It therefore attracts broad interest from multidisciplinary academic research and industry applications.

Because of the multiscale and complex nature of porous media, it is difficult to accurately characterize the microstructures and macroscopic morphology with

Euclidean geometry. The behavior of scaling phenomena assesses that all the scales within some range are equally important and the signals at different scales are related to each other. Since Mandelbrot (1982) proposed fractal geometry, it has been applied as a descriptive tool for porous media (Katz and Thompson, 1985; Sahimi, 2011; Hunt and Ewing et al., 2014; Dimri, 2016). The statistical self-similarity has been found in pore-scale and laboratory-scale as well as field-scale porous materials, and fractal geometry has been accordingly adopted to characterize pore and particle size distribution, pore area, solid mass, tortuous flow path, rough surface etc. in porous media. Fractal theory provides a promising framework for addressing the complexity and multiple scales of porous media, and the role of fractal theory in understanding the heat and mass transfer in porous media has opened a new branch of science and engineering (Yu and Xu et al., 2014).

A fractal can be defined as a geometrical shape with self-similarity independent of scale. It should be noted that most natural objects that are self-similar or self-affine have that property only in a statistical sense. That is, certain property (measured quantity) of a fractal object follows power-law relation or long-term dependence on another property of the same object. Mandelbrot (1982) argued that the unique property of fractal objects M is that they are independent of the unit of measurement (scale L) and follow the scaling law in the form of:

$$M(L) \sim L^{D_f} \tag{3.1}$$

A fractal object can be characterized by the fractal dimension (FD) D_f (typically non-integer) less than the Euclidean dimension it is embedded in. The fractal dimension can capture what is lost in traditional geometrical representation of form, which is usually estimated as the slope of a linear fit of the data on a log-log plot of a measure against the scale.

Pore and particle size distribution (PSD) form one of the most common descriptors in the field of porous media, and have been widely used to predict physical properties such as porosity, saturation, diffusivity, permeability, heat conduction, etc. Yu and Cheng (2002a) presented the cumulative size distribution in porous media as:

$$N(\varepsilon \geq \lambda) = (\lambda_{\max} / \lambda)^{D_f} \tag{3.2}$$

where N is the number of pores (or particles), ε and λ are the measure length scale and size respectively, the subscript max denotes the maximum size. Two kinds of fractal models of porous media, pore and particle (mass) fractal approaches, can be characterized with this scaling law (Xu, 2015). The probability density function (PDF) for pore-size distribution can be consequently obtained as:

$$f(\lambda) = D_f \lambda_{\min}^{D_f} \lambda^{-(D_f+1)} \tag{3.3}$$

where the subscript min represents the minimum pore size. Since the probability density function should meet the normalization condition, $(\lambda_{\min} / \lambda_{\max})^{D_f} = 0$ should be satisfied. Generally, the condition $\lambda_{\min} \ll \lambda_{\max}$ can be used as a criterion

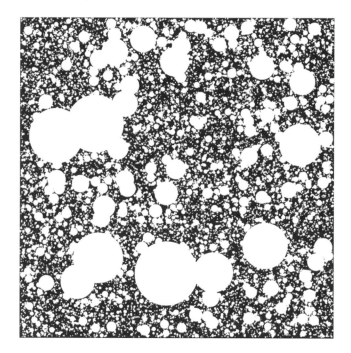

FIGURE 3.2 2D fractal porous media by Monte Carlo method (white part represents pores).

for fractal analysis of porous media. The cumulative distribution function (CDF) for pore-size distribution can be found as $F(\lambda) = 1 - (\lambda_{min} / \lambda)^{D_f}$, and the total number of pores is $N_t = (\lambda_{max} / \lambda_{min})^{D_f}$. Accordingly, the random pore sizes with fractal distribution can be generated by Monte Carlo method (Yu and Zou et al., 2005; Xu and Yu et al., 2013) via $\lambda = \lambda_{min} / (1 - \xi)^{1/D_f}$, where ξ is a set of random numbers of 0~1. Figure 3.2 shows a two-dimensional porous sample generated by fractal Monte Carlo simulation, where pore fractal dimension $D_f = 1.85$, porosity $\phi = 0.50$ and ratio of minimum pore diameter to maximum pore diameter $\lambda_{min} / \lambda_{max} = 10^{-2}$ were adopted.

The tortuous flow path (capillary) in porous media can be also characterized by a statistical fractal scaling law (Wheatcraft and Tyler, 1988; Yu and Cheng, 2002a):

$$L_t(\varepsilon) = \varepsilon^{1-D_T} L_0^{D_T} \tag{3.4}$$

where L_t and L_0 are the actual and straight length of the flow path and ε is the length scale of measurement (or capillary size). Tortuosity fractal dimension D_T corresponds to the tortuousness of the flow path.

The topology structure of capillaries or fractures in porous media may be characterized by a fractal tree-like network (Xu et al. 2016). As shown in Figure 3.3, the structure of a tree-like network is such that each successive daughter branch has a diameter smaller than that from which it originates. If the scale factors are

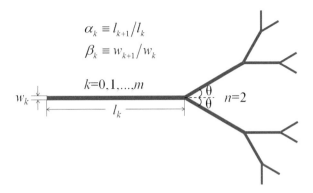

$$\alpha_k \equiv l_{k+1}/l_k$$
$$\beta_k \equiv w_{k+1}/w_k$$

$$k = 0, 1, \ldots, m$$

$$n = 2$$

FIGURE 3.3 A typical fractal tree-like network with symmetrical and dichotomous structure ($n = 2$).

independent of branching level k for the tree-like network with self-similar fractal property, then the similarity fractal dimensions for branch length and width can be defined as follow:

$$n = \alpha^{-D_l} = \beta^{-D_w} \tag{3.5}$$

where D_l and D_w are similarity fractal dimensions for length and width distributions, respectively.

The rough surface profiles always indicate a self-affine property, which means that surface roughness has different scaling factors in different directions (Mandelbrot, 1982; Sahimi, 2011). The surface profile R along x direction can be represented by the Weierstrass-Mandelbrot (W-M) function:

$$R(x) = G^{(D_R-1)} \sum_{n=n_1}^{\infty} \frac{\cos(2\pi\gamma^n x)}{\gamma^{(2-D_R)n}} \tag{3.6}$$

where G is a scaling constant, $1 < D_R < 2$ is the self-affine fractal dimension, $\gamma > 1$ is the scaling parameter for determining the spectral density, γ^n is the spatial frequency of the profile and n_1 is used to specify the low cut-off frequency with sample length L by $\gamma^{n_1} = 1/L$.

With the above fractal scaling laws, the structural parameters such as porosity, cross-sectional area (bulk volume), tortuosity, saturation etc. can be calculated accordingly. The porosity of fractal porous media can be determined by the pore fractal dimension (Yu and Cheng, 2002a):

$$\phi = (\lambda_{min}/\lambda_{max})^{d_E-D_f} \tag{3.7}$$

where d_E is the Euclidean dimension and $d_E = 2$ and $d_E = 3$ in two- and three-dimensional spaces, respectively. Then, the fractal dimension can be determined by $D_f = d_E - \ln\phi / \ln(\lambda_{min}/\lambda_{max})$ accordingly.

Assuming spherical pores, the cross-sectional area and bulk volume of a representative volume element (RVE) can be respectively calculated by:

$$A = \frac{\pi D_{f2} \lambda_{max}^2}{4(2 - D_{f2})} \frac{1 - \phi_s}{\phi_s} \tag{3.8a}$$

$$V = \frac{\pi D_{f3} \lambda_{max}^3}{6(3 - D_{f3})} \frac{1 - \phi_v}{\phi_v} \tag{3.8b}$$

where ϕ_s and ϕ_v are surface porosity and volume porosity, D_{f2} and D_{f3} are fractal dimensions in 2D and 3D spaces, respectively. And also, the average, maximum and minimum pore sizes can be calculated with the fractal probability density function (Yu and Xu et al., 2014).

If hydraulic tortuosity τ is defined as the ratio of actual flow-path length L_t to the geometrical length of a porous medium in the direction of macroscopic flux L_0, it can be expressed as:

$$\tau = (L_0 / \varepsilon)^{D_T - 1} \tag{3.9}$$

When multiphase fluids are concerned in fractal porous media, the wetting fluid saturation can be determined by (Xu and Qiu et al., 2013):

$$S_w = \frac{\left[4\sigma \cos\theta / (P_c \lambda_{max}) \right]^{2 - D_f} - \phi}{1 - \phi} \tag{3.10}$$

where σ is the surface tension of wetting phase, θ is the contact angle between liquid and solid phase and P_c is the effective capillary pressure. If both wetting and nonwetting phases indicate self-similar fractal scaling law, their fractal dimensions are (Yu and Xu et al., 2014):

$$D_{f,w} = d_E - \frac{\ln(S_w \phi)}{\ln(\lambda_{min} / \lambda_{max})} \tag{3.11a}$$

$$D_{f,nw} = d_E - \frac{\ln[(1 - S_w)\phi]}{\ln(\lambda_{min} / \lambda_{max})} \tag{3.11b}$$

The statistical fractal scaling laws can provide fine quantitative characterization of geometrically structural properties for estimating the heat and mass transfer properties of porous materials, understanding drying mechanisms and designing drying process.

3.2 FRACTAL MODELS FOR HEAT AND MASS TRANSFER

Since drying processes are directly dependent on the accurate prediction of transport properties of porous media, a comprehensive understanding of transport physics in porous media is of great commercial, environmental and scientific interest. Over the years, many empirical and semi-empirical as well as theoretical models have been developed for drying of porous media. Due to the multiscale nature of drying porous media, mesoscopic (pore-scale) models should be initiated at the first scale of interest (Prat, 2002). Accordingly, the macroscopic transport properties can be envisioned as an upscaling problem. During the drying process of porous media, the multiphysics fields including moisture content, temperature, fluid velocity, pressure, stress etc. can be determined according to suitable physical laws at pore level. The macroscopic transport equations can be obtained by averaging the pore-scale transport equations, which can be solved by suitable theoretical or numerical methods with specified boundary conditions. Therefore, pore-scale modeling of heat and mass transfer are needed for understanding drying dynamics and designing drying processes (Figure 3.4).

Recently, nanoporous materials open up new possibilities for various technologies, while drying technology plays an important role in synthesis and application of nanoporous media (Pakowski, 2004). The molecular dynamics and microscale effects such as slippage effect, capillary effect, Knudsen effect, Kelvin effect etc. may influence heat and mass transfer in nanoscale porous materials, which brings challenges for modeling drying of porous media (Thiery and Rodts et al., 2017; Yuan and Zhang et al., 2018). With scale invariant property, fractal geometry is able to quantify multiscale porous media down to the nanoscale level during the drying process. Thus, pore-scale physical and mathematical models for heat transfer, gas diffusion and fluid flow are presented by employing fractal geometry. And the effective

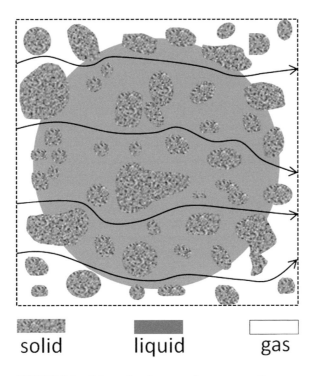

FIGURE 3.4 Schematic of pore-scale porous media.

transport parameters including thermal conductivity, diffusion coefficient, absolute permeability and relative permeability are proposed accordingly.

3.2.1 HEAT TRANSFER

Thermal conductivity and thermal dispersion conductivity are key parameters for heat transfer in porous media, which are very important for drying process of porous media. Three methods, Fourier's law models, Ohm's law models and empirical models, are usually used to predict thermal conductivity. Various experimental and numerical works have been conducted to explore the correlations for the effective thermal conductivity and thermal dispersion conductivity (Zhang and Wang, 2017). For example, Abu-Hamdeh et al. (2001) proposed an empirical formula for the effective thermal conductivity of sandy and clay soil via experimental measurement, which is a function of the density of porous media and water content. Adler and Thovert (1998) argued that thermal conductivity could be expressed in the form of Archie's law with porosity according to the numerical results obtained by solving the Laplace equation. Many theoretical models are also reported to predict thermal conductivity of porous media in literature. For example, Wiener (1912) presented a serial and parallel model of solid, liquid and gas phases for thermal conductivity:

$$k_{e,s} = \left[\sum (\phi_\alpha / k_\alpha) \right]^{-1}$$

(3.12a)

$$k_{e,p} = \sum \phi_\alpha k_\alpha$$

(3.12b)

where k and ϕ are respectively the thermal conductivity and volume fraction, subscript α denotes α phase. The effective thermal conductivity of serial and parallel models $k_{e,s}$ and $k_{e,p}$ present the upper and lower bounds of thermal conductivity of porous media.

Johansen (1975) proposed a geometric mean model to predict the thermal conductivity. The thermal conductivity for unsaturated porous media consists of solid, liquid and gas phases that can be expressed as:

$$k_e = k_s^{1-\phi} k_w^{\phi S_w} k_g^{\phi(1-S_w)}$$

(3.13)

where the subscripts e, s, w, g represent effective, solid, water and gas, respectively; ϕ is the porosity, and S_w is the saturation. Interested readers can refer to the book by Kaviany (1995) and review paper by Zhang and Wang (2017).

In order to understand the effect of compositional and environmental factors on the thermal conductivity of porous media, three fractal models for thermal conductivity of porous media at pore scale are introduced in the following section. Figures 3.5(a)-(c) show the fractal particle chain model, fractal capillary model and fractal Sierpinski model, respectively.

Based on fractal model for solid phase, Yu and Cheng (2002b) presented a fractal particle chain model of thermal conductivity. As shown in Figure 3.5(a), a porous medium is assumed to be composed by parallel mixed chains constituted by

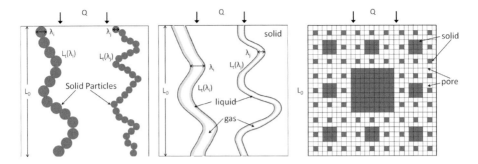

FIGURE 3.5 Pore-scale fractal models for thermal conductivity: (a) fractal particle chain model, (b) fractal capillary model, (c) fractal Sierpinski model.

touching particles and non-touching particles. Thus, the thermal resistance for one-dimensional heat conduction can be expressed as:

$$1 / R_t = 1 / R_{nt} + 1 / R_{mc} \tag{3.14}$$

where the subscripts t, nt and mc represent total, non-touching particles and mixed chains of touching particles, respectively. The effective thermal conductivity for non-touching particles $k_{e,n}$ can be calculated by (Hsu et al., 1995):

$$k_{e,n} = k_f \left(1 - \phi'\right) + \frac{k_f \phi'}{1 + \left(k_s / k_f - 1\right)\phi'} \tag{3.15}$$

where k is thermal conductivity, $\phi' = \sqrt{1 - \phi}$ and the subscripts f and s denote fluid and solid phase, respectively. The particle size distribution and tortuous chains are assumed to indicate statistical fractal scaling laws, which can be respectively characterized by Eqs. (3.3) and (3.4). Thus, the effective thermal conductivity for touching particle chains can be found with a cube particle model (Figure 3.6).

$$k_{e,mc} = k_f \frac{D_f}{1 + D_T - D_f} \frac{\lambda_{\max}^{1+D_T}}{A L_0^{D_T - 1}} \left[\frac{\gamma_a / \phi'}{\left(k_f / k_s\right)\phi' + 2\left(1 - \phi'\right)} + \frac{1 - \gamma_a}{\gamma_c^2 \phi'\left(k_f / k_s - 1\right) / \gamma_a^2 + 1} \right]^{-1} \tag{3.16}$$

where $\gamma_a = l / a$ and $\gamma_c = c / l$ represent the geometric length scale ratio and contact length scale ratio in the cube particle model. According to Eq. (3.14), the effective permeability can be obtained as:

$$k_e = \frac{A_n}{A} k_f \left[\left(1 - \phi'\right) + \frac{\phi'}{1 + \left(k_s / k_f - 1\right)\phi'} \right] + \left(1 - \frac{A_n}{A}\right) k_f \frac{D_f}{1 + D_T - D_f} \frac{\lambda_{\max}^{1+D_T}}{A L_0^{D_T - 1}}$$

$$\left[\frac{\gamma_a / \phi'}{\left(k_f / k_s\right)\phi' + 2\left(1 - \phi'\right)} + \frac{1 - \gamma_a}{\gamma_c^2 \phi'\left(k_f / k_s - 1\right) / \gamma_a^2 + 1} \right]^{-1} \tag{3.17}$$

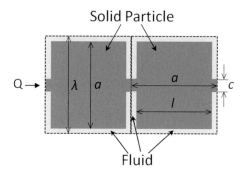

FIGURE 3.6 A cube particle model for touching particle chain.

where A is the total area of a representative cross section and A_n is an equivalent area of a cross section having the same porosity as the non-touching particles.

Based on the fractal model for pore phase, Kou and Liu et al. (2009) proposed a fractal model for effective thermal conductivity with fractal capillary model (Figure 3.5(b)). A porous medium with various pore sizes can be considered as a bundle of tortuous capillary, and the capillary size and length are assumed to follow the statistical fractal scaling laws (Eqs. (3.3) and (3.4)). Thus, the effective thermal conductivity can be calculated by parallel model of solid, water and gas phases.

$$k_e = \frac{1}{R_t}\frac{L_0}{A} = \frac{L_0}{A}\left(\frac{1}{R_s} + \frac{1}{R_w} + \frac{1}{R_g}\right) \tag{3.18}$$

where k and R are thermal conductivity and resistance, the subscripts e, t, s, w and g represent effective, total, solid, water and gas, respectively. The thermal resistance for the solid phase can be calculated by:

$$R_s = \frac{L_0}{(1-\phi)Ak_s} \tag{3.19}$$

A simplified model for the cross section of a capillary partially filled with water and gas is shown in Figure 3.5(b). Both water and gas phases indicate fractal scaling laws, which can be characterized by Eqs. (3.11a) and (3.11b) respectively. Therefore, the total thermal resistance for water (wetting) and gas (nonwetting) phases can be expressed by:

$$\frac{1}{R_w} = \int_{\lambda_{\min,w}}^{\lambda_{\max,w}} \frac{\pi\lambda^2 k_w}{4L_t(\lambda)} f(\lambda)N_t d\lambda \tag{3.20a}$$

$$\frac{1}{R_g} = \int_{\lambda_{\min,g}}^{\lambda_{\max,g}} \frac{\pi\lambda^2 k_g}{4L_t(\lambda)} f(\lambda)N_t d\lambda \tag{3.20b}$$

where the maximum and minimum diameters for water and gas phases are respectively:

$$\lambda_{max,w} = \lambda_{max}\sqrt{S_w} \tag{3.21a}$$

$$\lambda_{min,w} = \lambda_{min}\sqrt{S_w} \tag{3.21b}$$

$$\lambda_{max,g} = \lambda_{max}\sqrt{1 - S_w} \tag{3.21c}$$

$$\lambda_{min,g} = \lambda_{min}\sqrt{1 - S_w} \tag{3.21d}$$

With the aid of Eqs. (3.11a), (3.11b), (3.19), (3.20a) and (3.20b), the effective thermal conductivity can be found by Eq. (3.18):

$$
\begin{aligned}
k_e = (1-\phi)k_s &+ \frac{(2-D_f)D_{f,w}\lambda_{max,w}^{D_T+1}}{D_f(D_T-D_{f,w}+1)\lambda_{max}^2 L_0^{D_T-1}}\left[1-\left(\frac{\lambda_{min,w}}{\lambda_{max,w}}\right)^{D_T-D_{f,w}+1}\right]\frac{\phi}{1-\phi}k_w \\
&+ \frac{(2-D_f)D_{f,g}\lambda_{max,g}^{D_T+1}}{D_f(D_T-D_{f,g}+1)\lambda_{max}^2 L_0^{D_T-1}}\left[1-\left(\frac{\lambda_{min,g}}{\lambda_{max,g}}\right)^{D_T-D_{f,g}+1}\right]\frac{\phi}{1-\phi}k_g
\end{aligned}
\tag{3.22}
$$

where D_f, $D_{f,w}$ and $D_{f,g}$ are pore, water phase and gas phase fractal dimensions, respectively. It can be found that this effective thermal conductivity can reduce to the parallel model (Eq. (3.12b)) as tortuosity fractal dimension $D_T = 1$ and saturation $S_w = 0$ (or $S_w = 1$).

Self-similar Sierpinski models have been frequently applied to model the structures of porous media. As shown in Figure 3.5(c), a periodic porous structure is assumed and the Sierpinski model is used to characterize a unit cell of porous media (Ma and Yu et al., 2003). As the thermal conductivity strongly depends on the thermal conductivities of components, porosity and contact resistance, it is assumed that the porous medium consists randomly distributed non-touching particles and self-similar distributed particles in contact with each other and having a thermal resistance. The self-similarly contacting particles is simulated by the Sierpinski carpet. As shown in Figure 3.5(c), the ratio between the sides of successively sized squares is 3. The porosity and fractal dimension of Sierpinski carpet are $\phi = (8/9)^{n+1}$ and $D_f = \ln 8/\ln 3 \approx 1.893$, where $n = 0,1,2,\ldots$ denotes the stage of Sierpinski carpet. When two portions are considered to be in parallel, the total thermal resistance can be calculated by:

$$\frac{1}{R_t} = \frac{1}{R_{nt}} + \frac{1}{R_{sc}} \tag{3.23}$$

where the subscripts t, nt and sc represent total, non-touching particles and Sierpinski carpet, respectively. The thermal resistance for the non-touching particles can be evaluated by Hsu and Cheng's model (Eq. (3.15)). And the thermal resistance for self-similarly contacting particles can be calculated according to thermal-electrical analogy. Figure 3.7 displays the thermal-electrical model for the 0-stage Sierpinski

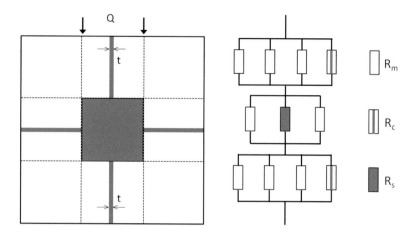

FIGURE 3.7 The thermal-electrical analogy for a 0-stage Sierpinski carpet: (a) thermal conductivity model, (b) thermal-electrical network.

carpet. The bars with width of t in Figure 3.7(a) represent the contact resistance between particles. The lateral contact thermal resistances can be neglected, and only the contact resistances along the heat flow direction are considered. Thus, the total resistance of the 0-stage Sierpinski carpet can be found by the thermal-electrical network shown in Figure 3.7(b), where R_m, R_c and R_s respectively denote the matrix thermal resistance (i.e., fluid), contact thermal resistance and solid thermal resistance. Then the thermal resistance of n-stage Sierpinski carpet can be obtained with recursive algorithm. The effective thermal conductivity of the porous media can be obtained with:

$$k_e = \left(\frac{1}{R_{nt}} + \frac{1}{R_{sc}} \right) \frac{L}{A} \qquad (3.24)$$

With a similar method, the effective thermal conductivity of three-phase porous media can be estimated also (Ma and Yu et al., 2004). It should be noted that the three-partition Sierpinski carpet is employed in above analysis; it can be extended to a generalized Sierpinski carpet (Feng and Yu et al., 2007a, 2007b).

For forced convection in porous media, thermal dispersion arising from the microscopic fluid velocity and temperature deviations from the average values play a remarkable role. Thermal dispersion is believed to enhance the heat transfer rate in porous media. And the effective thermal conductivity (k_e) can be taken as the summation of thermal conductivity of porous media with static fluid ($k_{e,f}$) and thermal dispersion conductivity (k_d).

$$k_e = k_{e,f} + k_d \qquad (3.25)$$

Experiments and empirical models indicate that the thermal dispersion conductivity depends on the flow velocity and microstructures in porous media. Yu and Li (2004) presented a theoretical model to predict the effective thermal dispersion conductivity with a fractal model. The tortuous streamline through porous media can be characterized by a fractal scaling law (Eq. (3.4)). Then, the difference between actual length and straight distance can be written as:

$$L' = L_t - L_0 = \lambda_{min}^{1-D_T} L_0^{D_T} - L_0 \tag{3.26}$$

Thus, the spatial deviation velocity along the direction perpendicular to the macroscopic flow direction can be found by differentiating Eq. (3.26) with respect to time t:

$$u' = \frac{dL'}{dt} = \left[D_T \left(\frac{L_0}{\lambda_{min}} \right)^{D_T-1} - 1 \right] u_0 \tag{3.27}$$

where $u_0 = dL_0 / dt$ is the straight velocity or volumetric average fluid phase velocity, and it equals to the ratio of Darcy velocity and porosity $u_0 = u_D / \phi$. If one-dimensional thermal dispersion is assumed, the thermal dispersion conductivity can be defined in the scalar form of

$$\rho_f C_{p,f} u'T' = -k_d \frac{d\langle T \rangle}{dy} \tag{3.28}$$

where ρ_f and $C_{p,f}$ are the density and specific heat at constant pressure of fluid and T and T' are temperature and spatial deviation temperature. Since the spatial deviation temperature along the y-direction can be written as $T' = -Cd_p d\langle T \rangle / dy$, the transverse thermal dispersion conductivity can be expressed as:

$$k_d = c\rho_f C_{p,f} d_p u' \tag{3.29}$$

where c is a fitting constant and d_p is particle diameter. With the aid of Eq. (3.27), the dimensionless thermal dispersion conductivity can be obtained as:

$$\frac{k_d}{k_f} = cP_e \left[D_T \left(\frac{L_0}{\lambda_{min}} \right)^{D_T-1} - 1 \right] \frac{1}{\phi} \tag{3.30}$$

where $P_e = \rho_f C_{p,f} d_p u_D / k_f$ is the Peclet number. The tortuosity fractal dimension can be determined by the box-counting method or by formulas of tortuosity. Actually, a simplified formula can be obtained via fitting the prediction by theoretical model and experimental data, which is believed to be more reasonable and accurate compared with empirical relationship from numerical and experimental work. For example, Yu and Li (2004) proposed a semi-empirical formula for thermal dispersion conductivity $k_d / k_f = c_D (1-\phi) P_e / \phi$, where fitting constant $c_D = 0.04$.

3.2.2 Gas Diffusion

Gas (air and water vapour) diffusion through a porous medium is very important for the drying process, which can be described by Fick's law. The effective gas diffusivity (diffusion coefficient) is one of the key parameters, which is commonly related to the bulk gas diffusivity D_b and porosity ϕ as well tortuosity τ.

$$D_e = D_b \phi / \tau \tag{3.31}$$

Since tortuosity can be also estimated by porosity, the effective diffusivity is often modeled as solely a function of porosity:

$$D_e = D_b f(\phi) \tag{3.32}$$

where f is the diffusibility. Various empirical and semi-empirical formulas have been proposed for the diffusibility as shown in Table 3.1 (Xu and Qiu et al., 2017).

Due to the multiscale nature of porous media, both molecular diffusion and Knudsen diffusion mechanisms may occur in pores. In the pores that are much larger than the gas molecule mean free path, the diffusion coefficient can be expressed by:

$$D_m = \frac{1}{3} lu \tag{3.33}$$

where l is the mean free path length and u is the molar-average velocity. The mean free path length and molar-average velocity can be respectively calculated by $l = \frac{k_B T}{\sqrt{2}\pi\delta^2 P}$ and $u = \sqrt{\frac{8k_B N_A T}{\pi M}}$, where k_B is the Boltzmann constant, T is the absolute temperature, δ is gas molecule diameter, P is the system pressure, N_A is Avogadro's number and M

TABLE 3.1

Available Formulas for the Diffusibility of Porous Media with Porosity

Diffusibility	Parameters	Porous Medium
$f(\phi) = \phi^m$	m is fitting constant	Particulate composite medium, Carbon paper/cloth
$f(\phi) = Ae^{B\phi}$	A and B are fitting constants	Porous gas diffusion layer
$f(\phi) = c\phi_g^a / \phi^b$	ϕ_g is air-filled porosity; a, b, and c are fitting constants	Soil
$f(\phi) = 1 - (1-\phi)^\beta$	β is fitting constant	Porous catalysts
$f(\phi) = 2\phi / (3-\phi)$		Isotropic spherical particle medium
$f(\phi) = \phi / \left[1 - \frac{1}{2}\ln\phi\right]$		Spherical particle bed
$f(\phi) = \phi\left(\dfrac{\phi - \phi_p}{1 - \phi_p}\right)^\alpha$	ϕ_p is percolation threshold, α is fitting constant	Fibrous media
$f(\phi) = 1 - q(\phi)\left[\dfrac{3(1-\phi)}{3-\phi}\right]$	$q(\phi)$ is empirical function of porosity	Fibrous gas diffusion layers

is the molecular weight. Gas molecules will collide with walls more frequently than with each other in the pores that are much smaller the mean free path. The Knudsen diffusivity can be expressed as:

$$D_k = \frac{1}{3}\lambda u \tag{3.34}$$

where λ is the pore diameter. The Knudsen number, which is defined as the ratio of the gas molecule's mean free path to the diameter $Kn = l / \lambda$, can be used as an indicator to evaluate the effect of Knudsen diffusion.

A fractal capillary bundle model as shown in Figure 3.8 is employed to derive the effective diffusion coefficient of porous media (Xu and Qiu et al., 2017). The capillary diameter and tortuous length are assumed to obey the statistical fractal scaling laws Eqs. (3.3) and (3.4), respectively. Then, the gas diffusion flux through a single tortuous capillary with diameter of λ can be expressed as:

$$j(\lambda) = \frac{\pi\lambda^2}{4} D \frac{\Delta C}{L_t} \tag{3.35}$$

where $\Delta C / L_t$ is the concentration gradient along the tortuous capillary. The gas diffusion coefficient D depends on the capillary size. As the value of the Knudsen

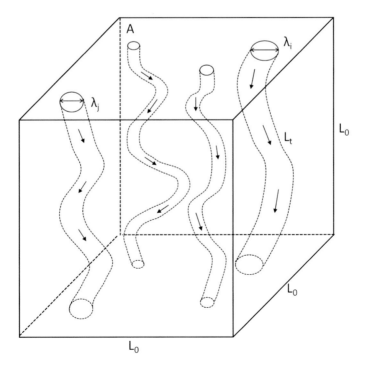

FIGURE 3.8 A schematic graph of fractal capillary bundle model for porous media.

number is smaller than 1 in the porous media, the gas diffusion coefficient takes the value of the molecular diffusivity (Eq. (3.33)). Accordingly, the total gas diffusion flux through a RVE can be obtained by integrating the individual diffusion flux over the entire range of pore sizes.

$$J = \int_{\lambda_{min}}^{\lambda_{max}} j(\lambda) f(\lambda) N_t d\lambda = \frac{\pi u l}{12} \frac{D_f \lambda_{max}^{1+D_T}}{1+D_T-D_f} \frac{\Delta C}{L_0^{D_T}} \left[1 - \left(\frac{\lambda_{min}}{\lambda_{max}} \right)^{1+D_T-D_f} \right] \qquad (3.36)$$

According to Fick's law, $J = D_e A \Delta C / L_0$, the effective gas diffusivity of the porous medium can be obtained as:

$$D_{em} = \frac{1}{3} l u C_D \left(\frac{\phi}{1-\phi} \right)^{(1+D_T)/2} \left[1 - \phi^{(1+D_T-D_f)/(2-D_f)} \right] \qquad (3.37)$$

where the fractal coefficient is $C_D = (\pi D_f / 4)^{(1-D_T)/2} (2-D_f)^{(1+D_T)/2} / (1+D_T - D_f)$, the value of which is 1.0 as the tortuosity fractal dimension is $D_T = 1.0$.

If molecular diffusion and Knudsen diffusion mechanisms coexist in multiscale porous media, the diffusion coefficient defined in Eq. (3.35) takes the value of molecular diffusion coefficient D_m as $Kn < 1$ and the Knudsen diffusion coefficient D_k for $Kn > 1$. The effective gas diffusivity can be found with a similar method.

$$D_e = \frac{1}{3} D_m C_D \left(\frac{\phi}{1-\phi} \right)^{(1+D_T)/2} \left[1 - \frac{Kn_{min}^{1+D_T-D_f}}{2+D_T-D_f} \left(1 - \frac{1+D_T-D_f}{Kn_{max}^{2+D_T-D_f}} \right) \right] \qquad (3.38)$$

where $Kn_{min} = l / \lambda_{max}$ and $Kn_{max} = l / \lambda_{min}$ are the minimum and maximum Knudsen numbers in the porous media, respectively. It can be found that the effective gas diffusivity is determined by the bulk diffusion and pore structure as well as the Knudsen number.

3.2.3 FLUID FLOW

The basic law governing fluid flow through porous media is Darcy's law. The flow rate through a cross section A can be expressed as:

$$q = \frac{K}{\mu} A \frac{\partial P}{\partial x} \qquad (3.39)$$

where μ is the dynamic viscosity of fluid, $\partial P / \partial x$ is the pressure gradient in the flow direction and permeability K depends on the geometry of porous medium but is independent of nature of fluid. Actually, permeability is in general a second-order tensor in three dimensions. The permeability in Eq. (3.39) is taken as a scalar for the case of an isotropic medium. The values of permeability for different kinds of porous media vary widely. Although a great deal of work has been conducted to evaluate and predict the permeability of porous media for more than a century, a generalized theoretical expression for permeability is still needed. The classical permeability-porosity relation, the Kozeny-Carman equation, is the most common method, which is also widely used as the starting point for many other permeability models.

$$K = \frac{\phi^3}{c_k (1 - \phi)^2 S^2} \tag{3.40}$$

where S denote the specific surface area based on the solid volume and c_k is Kozeny-Carman constant. It should be pointed out that Kozeny-Carman constant is actually an empirical constant, which was proved to be not a constant and may be related to porosity (Xu and Yu, 2008).

Since Mandelbrot (1982) proposed fractal geometry, fractal theory has been successfully applied to characterize the complex microstructures and determine the permeability of porous media. For example, various self-similar Sierpinski carpets and Menger sponges have been adopted to model the structure of porous media, and a computational fluid dynamics method can be performed to study the fluid flow and calculate the permeability of porous media. Figure 3.9 shows the velocity contour and flow pathlines in a 5-stage Sierpinski carpet with circular particles.

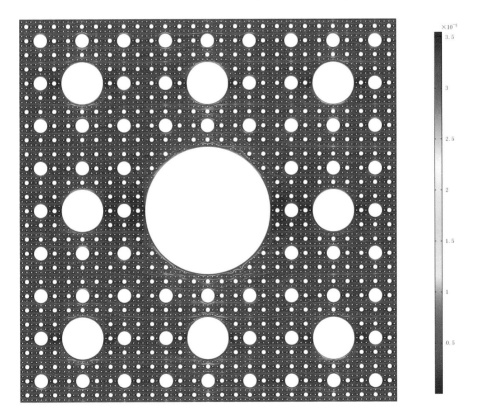

FIGURE 3.9 The velocity contour and flow pathlines in a 5-stage Sierpinski carpet with circle particles.

In this section, a fractal capillary bundle model is introduced as this model can provide understanding of physical mechanisms and can be taken as a predictive model. The fractal capillary bundle model (Figure 3.8) is similar to that in section 3.2.2, and the same fractal scaling laws are applied here. According to the Hagen-Poiseuille equation, the flow rate in a pipe (capillary) as Knudsen number $Kn \leq 0.001$ can be written as:

$$q(\lambda) = \frac{\pi \lambda^4}{128\mu} \frac{\Delta P}{L_t(\lambda)} \tag{3.41}$$

where ΔP is the pressure drop along the capillary. The total flow rate over the entire range of pore sizes from minimum to maximum pore diameters can be obtained according to mass conservation.

$$Q = \int_{\lambda_{min}}^{\lambda_{max}} q(\lambda) f(\lambda) N_t d\lambda = \frac{\pi}{128\mu} \frac{D_f}{3 + D_T - D_f} \frac{\Delta P}{L_0^{D_T}} \lambda_{max}^{3+D_T} \tag{3.42}$$

Using Darcy's law, the permeability of a porous medium can be found as follows:

$$K = \frac{\pi}{128A} \frac{D_f}{3 + D_T - D_f} \frac{\lambda_{max}^{3+D_T}}{L_0^{D_T-1}} \tag{3.43a}$$

With the aid of Eqs. (3.7) and (3.8), it can be also expressed as:

$$K = C_f \lambda_{max}^2 \left(\frac{\phi}{1-\phi}\right)^{(1+D_T)/2} \tag{3.43b}$$

where $C_f = \frac{(\pi D_f)^{(1-D_T)/2}(2-D_f)^{(1+D_T)/2}}{2^{(6-D_T)}(3+D_T-D_f)}$ is a fractal factor determined by fractal dimensions. Accordingly, Xu and Yu (2008) presented an analytical expression for Kozeny-Carman constant.

$$c_k = \frac{1}{36C_f} \frac{\phi^{(3-D_T)/2}}{(1-\phi)^{(1-D_T)/2}} \tag{3.44}$$

For micro- and nanoscale porous media, gas slippage may indicate a significant effect on the gas flow as the gas molecule's mean free path is comparable to the pore size. Li and Xu et al. (2016) proposed a generalized form of effective permeability for all flow regimes including continuum ($Kn < 0.01$), slip flow ($0.01 < Kn < 0.1$), transition ($0.1 < Kn < 10$) and free molecular flow ($Kn > 10$) regimes.

$$K = C_f \lambda_{max}^2 (1 + \alpha Kn)\left(1 + \frac{4Kn}{1 - bKn}\right)\left(\frac{\phi}{1-\phi}\right)^{(1+D_T)/2} \tag{3.45}$$

where α is a dimensionless rarefied effect coefficient and b is the slip factor. In the slip flow regime, $\alpha = 0$ and $b = -1$ are recommended by Karniadakis and Beskok (2002), the analytical expression for the Klinkenberg's slippage factor can be found as:

$$b_K = \frac{\mu(3 + D_T - D_f)}{\lambda_{\max}(2 + D_T - D_f)}\left(\frac{32\pi R_g T}{M}\right)^{1/2} \tag{3.46}$$

For the unsaturated porous media, Yu and Li et al. (2003) proposed analytical expressions for the relative permeabilities for wetting and nonwetting phases with the assumption that both wetting and nonwetting phases follow the same fractal scaling laws (Eqs. (3.3) and (3.4)).

$$k_{rw} = \frac{K_w}{K} = \frac{3 + D_T - D_f}{3 + D_T - D_{f,w}}\frac{D_{f,w}}{D_f}S_w^{(3+D_T)/2} \tag{3.47a}$$

$$k_{rg} = \frac{K_g}{K} = \frac{3 + D_T - D_f}{3 + D_T - D_{f,g}}\frac{D_{f,g}}{D_f}(1 - S_w)^{(3+D_T)/2} \tag{3.47b}$$

Xu and Qiu et al. (2013) presented another fractal model for unsaturated porous media with the assumption that the nonwetting fluid occupies all pores larger than a critical value at certain value of capillary pressure.

$$k_{rw} = \frac{K_w}{K} = S_w\left[(1 - \phi)S_w + \phi\right]^{(3+D_T-D_f)/(2-D_f)} \tag{3.48a}$$

$$k_{rg} = \frac{K_g}{K} = (1 - S_w)\left\{1 - \left[(1 - \phi)S_w + \phi\right]^{(3+D_T-D_f)/(2-D_f)}\right\} \tag{3.48b}$$

3.3 FRACTAL IN DRYING POROUS MEDIA

Nature provides a considerable variety of shapes and forms, from the simple to the complex. Generally, regular objects such as points, lines, surfaces and bodies can be described by Euclidean geometry with integer dimensions (0, 1, 2 and 3), the measures of which are invariant with respect to the unit of measurement used. Numerous irregular shapes can be characterized and quantified using the concept of fractal geometry, and the degree of geometrical irregularity can be measured by fractal dimension. The unique property of fractal objects is that they are independent of the unit of measurement and follow the scaling law in the form of Eq. (3.1). Introducing fractal theory into drying porous media can provide new insights to the heat and mass transfer mechanisms (Xu and Mujumdar et al., 2008). The application of fractal concepts to drying porous media can be categorized into two aspects, one is characterizing geometrical structures of porous material in drying with fractal geometry and the other is heat and mass transfer models for drying porous media developed with fractal theory.

3.3.1 MICROSTRUCTURE OF POROUS MEDIA

The microstructures of porous media such as the size, shape and connection of pores and particles etc. are important for evaluating macroscopic properties such as thermal conductivity, moisture diffusivity, fluid permeability etc. The experimental methods for characterization of the laboratory-scale microstructures include mercury porosimetry, sorption measurements and tomographic methods such as X-ray imaging, magnetic resonance imaging, computed tomography and three-dimension transmission electronmicroscopy etc. However, the proper interpretation of the data is not straightforward and requires careful modeling. Fractal geometry has become a new and powerful method to quantify the geometric and structural data and images of porous media during drying process.

Table 3.2 lists representative examples for characterizing the microstructures of porous media under drying with fractal geometry. Yano and Nagai (1989) performed fractal analysis on the surface structure of native potato starch, wheat flour and rice under freeze drying at −20°C and 20 Pa. Their results from nitrogen adsorption isotherm show that the fractal dimensions of particle size distribution for fresh potato, wheat and rice are 2.20, 2.34 and 2.29, while the fractal dimensions for the transformed food with ethanol are 3.0, 2.9 and 3.0, respectively. The fractal dimension for the freeze-dried potato is 2.7. They therefore concluded that transformation with ethanol and freeze drying can increase the fractal dimension. Reich et al. (1992) interpreted the small-angle X-ray scattering data for air-dried brown coal, and reported that surface fractal dimensions for wet/dry dark/light lithotypes are respectively 3.46, 2.60, 3.47 and 2.47, and that volume (mass) fractal dimension are 2.54, 3.40, 2.53 and 3.33. They argued that mass and surface fractal terms dominate the scattering from bed-moist and dried brown coal respectively. Nussinovitch et al. (2004) characterized the pore-size distribution on the surface of freeze-dried agar-texturized fruit at 33.3 Pa and −45°C with fractal theory. They got the average fractal dimension for blank, 10% orange and 10% banana dried cellular solids as 1.11, 1.21 and 1.15 using scanning electron microscopy and the box-counting method. Rahman et al. (2005) reported that fractal dimensions for pore-size distribution in fresh, air-dried apple products after 20 and 30 h of drying at 80°C were 2.71, 2.96 and 3.11 according to the mercury porosimetry experiments. It is found that the fractal dimension increases with the increase of drying time indicating formation of micro-pores on the surface during air drying. However, Pfeifer and Avnir (1983) indicates that the fractal dimension for pore-size distribution is in the range of 2 to 3. Kerdpiboon et al. (2007) and Kerdpiboon and Devahastin (2007) discussed the effect of drying time and drying method on the fractal dimension for cell walls of carrot and potato cubes with light microscope. The fractal dimension of carrot increases from 1.75 to 1.91 after 300 min hot air drying at 60°C and 0.5 m/s, while the fractal dimension for potato varies from 1.61 to 1.73 after 420 min hot air drying at 60°C and 1 m/s. Linear and non-linear changes of fractal dimensions were respectively observed during and after the first falling drying rate period. During different drying techniques, the microstructural changes of carrot samples were quite different, but the fractal dimension was found to increase with drying time for both hot air dying and low-pressure superheated steam drying. Campos-Mendiola et al. (2007) calculated the fractal dimension of solid-gas

TABLE 3.2
Fractal Characterization of Microstructures of Porous Media in Drying

MicroStructure	Porous Media	Drying Technique	Drying Cond.	Measure	Fractal Dimension	References
Particle size distribution	Potato starch Wheat flour Rice	Freeze drying	−20°C 20 Pa	Nitrogen adsorption isotherm	• FD for fresh potato, wheat and rice: 2.20, 2.34, 2.29 • FD for potato, wheat and rice transformed with ethanol: 3.0, 2.9, 3.0 • FD for freeze-dried potato: 2.7	Yano and Nagai, 1989
Pore structure	Brown coal	Air drying	/	Small-angle X-ray scattering	• Surface FD for wet/dry dark/light lithotypes: 3.46, 2.60, 3.47, 2.47 • Volume FD are 2.54, 3.40, 2.53, 3.33	Reich et al., 1992
Pore-size distribution	Agar-texturized fruit	Freeze drying	−45°C 33.3 Pa	Scanning electron microscopy	• Average FD for blank, 10% orange, 10% banana: 1.11, 1.21 and 1.15	Nussinovitch et al., 2004
Pore-size distribution	Apple	Air drying	80°C	Mercury porosimetry	• FD for fresh, air-dried apple products after 20 and 30 h drying: 2.71, 2.96, 3.11	Rahman et al., 2005
Cell wall	Carrot Potato	Air drying Low-pressure superheated steam drying	60–80°C 0.5, 1 m/s	Light microscope	• FD of carrot increases from 1.75 to 1.91 after 300 min hot air drying at 60°C and 0.5 m/s • FD of potato varies from 1.61 to 1.73 after 420 min hot air drying at 60°C and 1 m/s	Kerdpiboon and Devahastin, 2007a Kerdpiboon et al., 2007
Solid-gas interface	Potato	Air drying	40–70°C 1–2.5m/s	Stereoscopic microscope	• FD for drying time 0, 10, 15, 30, 40 and 100 min: 1.067, 1.076, 1.071, 1.050, 1.143, 1.101	Campos-Mendiola et al., 2007
Particle and Pore	Silica aerogels	Supercritical drying	Supercritical CO_2	Small-angle X-ray scattering and nitrogen adsorption	• Mass FD for wet gels is around 2.25 • Mass and surface FDs for aerogels are around 2.4 and 2.6	Perissinotto et al., 2015

interfaces for potato during hot air drying at four temperature (40, 50, 60 and 70°C) and air velocities (1, 1.5, 2.0 and 2.5 m/s) with stereoscopic microscopic images. The fractal dimension for drying time 0, 10, 15, 30, 40 and 100 min are 1.067, 1.076, 1.071, 1.050, 1.143, 1.101, respectively. They argued that the tendency of the fractal dimension of the interfaces is similar to the one found for lateral projected area, which may indicated that inner structures composed of solid-gas matrixes are projected towards the surface in the form of fractal solid-gas interfaces. Perissinotto et al. (2015) found that the mass and surface fractal dimension for supercritical dried silica aerogels are around 2.4 and 2.6, as inferred from small-angle X-ray scattering and nitrogen adsorption data. Aerogels presented most of the mass-fractal characteristics of the original wet gels at large length scales.

It can be found that fractal scaling laws can be used to characterize the statistical properties of pore, particle, cell and solid-gas interface in porous media during drying, which provides quantitative data suitable to modeling and improve the understanding of drying mechanisms. And also, fractal dimensions can be correlated with the properties of dried porous products, which can be applied to assist designing drying process.

3.3.2 MACROSCOPIC MORPHOLOGY

Both the macroscopic and microscopic structures of porous material in the drying process change significantly during the drying process due to dehydration, which consequently affects the macroscopic properties of dried materials. It is difficult to describe the macroscopic morphology of porous media such as rough surface and irregular shape with traditional methods. Also, the damage of microstructures can accordingly change the macroscopic morphology and make it more complex in the drying process. Therefore, fractal geometry has been successfully applied to characterize the macroscopic morphology such as rough surface and cracks in the drying process, which may shed light on drying properties and heat and mass transfer mechanisms.

Winslow et al. (1995) determined the fractal type and dimensions of hydrated Portland cement pastes over a range of length scales via small-angle X-ray scattering experiments. Their results indicate that the larger-scale (200-1500Å) geometry at most degrees of water saturation is that of a rough surface fractal, while the smaller-scale (30-200 Å) is a mass fractal in saturated pastes. Chanona et al. (2003) studied the influence of air drying time and condition on the fractal dimension for surface images of carbohydrate-based food. They reported that the fractal dimensions are 2.1666 and 2.2593 at 0 and 150 min after drying under 45°C and 1 m/s, while the fractal dimensions after 300 min drying are respectively 2.3703 and 2.4520 at different drying conditions (45°C, 1 m/s and 60°C, 3 m/s). It is indicated that fractal dimension tends to be higher for longer drying processes, and higher temperatures and airflows gave the highest values of fractal dimension. The surface temperature distribution also exhibits a non-linear behavior, which can be characterized by a fractal dimension. Balankin et al. (2006) proposed that the rough configurations of dried paper sheets follow self-similar scaling laws. The fractal dimensions of each original profile determined by the box-counting, divider and perimeter-area methods are almost the same. The surface fractal dimensions for Filtro paper with open medium and closed porosity were evaluated to be 2.24, 2.13 and 2.09, respectively.

Cracking caused by rapid drying at high temperatures is an important index of the product quality; however, it is an interior defect within products and it is difficult to describe and understand the picture of stress cracks completely (Xu et al., 2009). Tao et al. (2005) characterized the rugged boundaries and cross sections with Koch island and Sierpinski carpet according to the X-ray microcomputerized tomography scanned data of sludge crack, and obtained averaged boundary and cross-sectional fractal dimensions of 1.4 and 2, respectively. Wang and Li (2009) studied the fractal dimensions according to the drying-rate curve to qualitatively describe the shrinkage and cracks of sludge and understand the internal dynamics of the drying system. The monofractaldimensions for cylindrical drying sludge with diameter of 3, 6 and 18 mm at 120°C are 1.7187, 1.7151 and 1.8479, respectively. The multifractal dimension is a monotone function of factorial scaling exponent of probability measure. Smyth et al. (2014) reported that the fractal dimension for the rough surface of articular cartilage during drying indicates self-similarity over the viewing scale, which remains relatively unchanged although the average roughness of a cartilage surface changes rapidly with drying time.

3.3.3 Fractal Model for Drying

Drying as a process combining heat and mass transfer contains many possible physical mechanisms such as multiphase flow, diffusion, heat convection and heat conduction etc. As mentioned above, fractal theory presents an effective tool to characterize the microscopic and macroscopic structures of porous media; it is a possible way to theoretically analyze heat and mass transfer mechanisms and identify the complex multiphysics in the drying process. As discussed in section 3.2, the effective transport parameters, including thermal conductivity, diffusion coefficient, absolute permeability and relative permeability, can be derived with pore-scale fractal models. Also, several effective fractal models for drying porous media such as fractal percolation theory, fractal Brownian motion and fractal diffusion model as well as fractal fracture model etc. have been proposed.

The drying process can be simulated with the invasion percolation model. It has been shown that drying front and drainage front are similar to fractal object; the fractal dimension of drainage front is consistent with the fractal dimension of the external perimeter of the percolation cluster (Prat,1995; Tsimpanogiannis and Yortsos, 1999). Doulia et al. (2000) stated that the contrail of Brownian motion during molecular transport is fractal and therefore proposed several equations to describe molecular and convective transport through fractal structures and applied to chemical kinetics and transport in multiphase media. Mehrafarin and Faghihi (2001) reported that the falling-rate regime of the drying-rate curve of porous materials can be processed as a random-walk diffusion on a fractal structure representing the air-filled pore space, which yields anomalous diffusion for vapour particles. Okubo et al. (2004) studied the drying dissipative structures of aqueous solution of poly and the aqueous suspensions of monodisperse bentonite, and they found that the fractal dimension for the area increased from 1.2 to 1.6 as polymer concentration increases from 10^{-6} monoM to 10^{-2} monoM, and slightly increases as drying temperature increased. Zhu et al. (2004) applied self-similar fractal theory to study the irregular and rough profile of stress cracks of corn kernel and established four types of fractal propagating models of stress cracks (tergranule, transgranule, combination of tergranule and transgranule and bifurcation fracture model). Chai (2007) identified fractal features and mechanisms of drying sludge, and a parametric study of some control conditions on fractal dimension of structures was conducted. Yiotis et al. (2010) performed numerical simulations of isothermal drying on a 2D regular square network of spherical pores connected through cylindrical throats in order to provide a better understanding of the structures of the drying patterns; the fractal dimensions for the invading gaseous phase and drying front perimeter are calculated to be 1.88 ± 0.03 and 1.34 ± 0.06 in the absence of gravity.

3.4   CONCLUDING REMARKS

Fractal geometry, with its ability to describe irregular objects that traditional Euclidean geometry fails to analyze, provides a new language for drying porous media. The main contents of this chapter are summarized as follows:

1. The basic concept of fractal geometry and fractal scaling laws for pore and particle size distribution, tortuous flow path, tree-like network and rough surface are introduced, and the formulas for porosity and tortuosity as well as saturation are also presented.
2. Pore-scale physical and mathematical models for heat transfer, gas diffusion and fluid flow are developed by the means of fractal theory, and the analytical expressions for thermal conductivity, diffusion coefficient, absolute permeability and relative permeability are derived accordingly.
3. Applications of fractal geometry in characterization of porous media during the drying process such as pore and particle structure, pore and particle size distribution, cell wall, solid-gas interface, rough surface, crack and drying-rate curve etc. are reviewed, and several effective fractal models for drying porous media such as fractal percolation theory, fractal Brownian motion and fractal diffusion model as well as fractal fracture model etc. are addressed.

It is clear that fractal theory has advantages for identifying the complex structure of porous material during the drying process and understanding the heat and mass transfer mechanisms in drying porous media. However, there is actually some urgent and challenging research needed in this field, which are as follows:

1. Some confusion and limitations in the application of fractal theory in drying porous media are found; the essential principles for fractal drying should be established.
2. More universal relationships between the drying process (method, drying condition, time etc.) and fractal characteristics need to be developed.
3. Correlating the fractal dimension of dried porous materials and its variations with the qualities of products and using the fractal dimension to choose and design the drying process can be a very interesting and challenging topic for future investigation.

ACKNOWLEDGEMENTS

This work was jointly supported by the National Natural Science Foundation of China (No. 51876196) and Zhejiang Provincial Natural Science Foundation of China (No. LR19E060001).

REFERENCES

Abu-Hamdeh, N.H., Khdair, A.I., and Reeder R.C. 2001. A comparison of two methods used to evaluate thermal conductivity for some soils. *International Journal of Heat and Mass Transfer* 44: 1073–1078.

Adler, P.M., and Thovert, J. F. 1998. Real porous media: local geometry and macroscopic properties. *Applied Mechanics Reviews* 51: 537–585.

Balankin, A.S., Morales, D., Susarrey, O., et al. 2006. Self-similar roughening of drying wet paper. *Physical Review E* 73: 065105.

Campos-Mendiola, R., Hernández-Sánchez, H., Chanona-Pérez, J.J., et al. 2007. Non-isotropic shrinkage and interfaces during convective drying of potato slabs within the frame of the systematic approach to food engineering systems (SAFES) methodology. *Journal of Food Engineering* 83: 285–292.

Chai, L.H. 2007. Statistical dynamic features of sludge drying systems. *International Journal of Thermal Science* 46: 802–811.

Chanona, P.J.J., Alamilla, B.L., Farrera, R.R.R., et al. 2003. Description of the convective air-drying of a food model by means of the fractal theory. *Food Science and Technology International* 9: 207–213.

Defraeye, T. 2014. Advanced computational modelling for drying processes-A review. *Applied Energy* 131: 323–344.

Dimri, V.P. 2016. *Fractal Solutions for Understanding Complex Systems in Earth Sciences.* Cham: Springer.

Doulia, D., Tzia, K., Gekas, V. 2000. A knowledge base for the apparent mass diffusion coefficient (D_{EFF}) of foods. *International Journal of Food Properties* 3: 1–14.

Feng, Y., Yu, B.M., Zou, M., Zhang, D. 2007a. A generalized model for the effective thermal conductivity of porous media based on self-similarity. *Journal of Physics D Applied Physics* 37: 3030–3040.

Feng, Y., Yu, B.M., Zou, M., Xu, P. 2007b. A generalized model for the effective thermal conductivity of unsaturated porous media based on self-similarity. *Journal of Porous Media* 10: 551–568.

Hsu, C. T., Cheng, P., Wong, K. W., 1995. A lumped-parameter model for stagnant thermal conductivity of spatially periodic porous media. *ASME Journal of Heat Transfer* 117: 264–269.

Hunt, A., Ewing R., Ghanbarian B. 2014. *Percolation Theory for Flow in Porous Media*, 3rd Ed. New York: Springer.

Johansen, O. 1975. Thermal conductivity of soils. PhD thesis, University of Trondheim.

Karniadakis, G.E., and Beskok, A., 2002. *Micro-flows, Fundamentals and Simulation.* New York: Spinger-Verlag.

Katz, A.J., and Thompson, A.H. 1985. Fractal sandstone pores: implications for conductivity and pore formation. *Physical Review Letters* 54: 1325–1328.

Kaviany, M. 1995. *Principles of Heat Transfer in Porous Media.* New York: Springer-Verlag.

Kerdpiboon, S., and Devahastin, S. 2007. Fractal characterization of some physical properties of a food product under various drying conditions. *Drying Technology* 25: 135–146.

Kerdpiboon, S., Devahastin, S., Kerr, W.L. 2007. Comparative fractal characterization of physical changes of different food products during drying. *Journal of Food Engineering* 83: 570–580.

Kou, J., Liu, Y., Wu, F., Fan, J., Lu, H., Xu, Y., 2009. Fractal analysis of effective thermal conductivity for three-phase (unsaturated) porous media. *Journal of Applied Physics* 106: 054905.

Kowalski, S.J., 2003. *Thermomechanics of Drying Processes*. Berlin: Springer.

Li, C., Xu, P., Qiu, S., Zhou, Y. 2016. The gas effective permeability of porous media with Klinkenberg effect. *Journal of Natural Gas Science and Engineering* 34: 534–540.

Ma, Y., Yu, B.M., Zhang, D., Zou, M. 2003. A self-similarity model for effective thermal conductivity of porous media. *Journal of Physics D Applied Physics* 36: 2157.

Ma, Y., Yu, B.M., Zhang, D., Zou, M. 2004. Fractal geometry model for effective thermal conductivity of three-phase porous media. *Journal of Applied Physics* 95: 6426–6434.

Mandelbrot, B.B. 1982. *The Fractal Geometry of Nature*. New York: Freeman.

Mehrafarin, M., and Faghihi, M. 2001. Random-walk diffusion and drying of porous media. *Physica A* 301: 163–168.

Mujumdar, A.S., and Passos, M.L. 2000. Innovation in drying technologies. In *Drying Technology in Agriculture and Food Science*, ed. A.S. Mujumdar, 291–310. Enfield: Science Publishers Inc.

Nussinovitch, A., Jaffe, N., Gililov, M. 2004. Fractal pore-size distribution freeze-dried agar-texturized fruit surface. *Food Hydrocolloids* 18: 825–835.

Okubo, T., Kanayama, S., Ogawa, H., Hibino, M., Kimura, K. 2004. Dissipative structures formed in the course of drying an aqueous solution of poly (allylamine hydrochloride) on a cover glass. *Colloid & Polymer Science* 282: 230–235.

Pakowski Z. 2004. Drying of nanoporous and nanostructured materials. In *Proceedings of the 14th International Drying Symposium*, 69–88. São Paulo, Brazil.

Perissinotto, A. P., Awano, C.M., Donatti, D.A., et al. 2015. Mass and surface fractal in supercritical dried silica aerogels prepared with additions of sodium dodecyl sulfate. *Langmuir* 31: 562–568

Pfeifer, P., and Avnir, D. 1983. Chemistry in noninteger dimensions between two and three. I. Fractal theory of heterogeneous surfaces. *The Journal of Chemical Physics* 79: 3558–3565.

Prat, M. 1995. Isothermal drying of non-hygroscopic capillary-porous materials as an invasion percolation process. *International Journal of Multiphase Flow* 21: 875–892.

Prat, M. 2002. Recent advances in pore-scale models for drying of porous media. *Chemical Engineering Journal* 86: 153–164.

Rahman, M.S., Al-Zakwani, I., Guizani, N. 2005. Pore formation in apple during air-drying as a function of temperature: porosity and pore-size distribution. *Journal of the Science of Food and Agriculture* 85: 979–989.

Reich, M.H., Snook, I.K., Wagenfeld, H.K. 1992. A fractal interpretation of the effect of drying on the pore structure of Victorian brown coal. *Fuel* 71: 669–672.

Sahimi M. 2011. *Flow and Transport in Porous Media and Fractured Rock: From Classical Methods to Modern Approaches*, 2nd Ed. New York: Wiley-VCH.

Smyth, P.A., Rifkin, R., Jackson, R.L., Hanson, R.R. 2014. The average roughness and fractal dimension of articular cartilage during drying. *Scanning* 36: 368–375.

Tao, T., Peng, X.F., Lee, D.J. 2005. Structure of crack in thermally dried sludge cake. *Drying Technology* 23: 1555–1568.

Thiery, J., Rodts, S., Weitz, D.A., Coussot, P. 2017. Drying regimes in homogeneous porous media from macro- to nanoscale. *Physical Review Fluids* 2: 074201.

Tsotsas, E., and Mujumdar, A.S. 2007. *Modern Drying Technology, Volume 1: Computational Tools at Different Scales*. Darmstadt: WILEY-VCH.

Tsimpanogiannis, I.N., and Yortsos, Y.C. 1999. Scaling theory of drying in porous media. *Physical Review E* 59: 4353–4365.

Wang, W., and Li, A. 2009. Multifractal characteristics of cylindrical sludge drying. *Frontiers Environmental Science Engineering in China* 3: 464–469.

Wheatcraft, S.W., and Tyler, S.W. 1988. An explanation of scale-dependent dispersivity in heterogeneous aquifers using concepts of fractal geometry. *Water Resource Research* 24: 566–578.

Wiener, O. 1912. Abhandl math-phys Kl KoniglSachsischenGes, 32: 509.

Winslow, D., Bukowski, J.M., Young, J.F. 1995.The fractal arrangement of hydrated cement paste. *Cement and Concrete Research* 25: 147–156.

Xu, P. 2015. A discussion on fractal models for transport physics of porous media. *Fractals* 23: 1530001.

Xu, P., Mujumdar, A.S., Yu, B.M. 2008. Fractal theory on drying: a review. *Drying Technology* 26: 1–11.

Xu, P., Mujumdar, A.S., Yu, B.M. 2009. Drying-induced cracks in thin film fabricated from colloidal dispersions. *Drying Technology* 27: 636–652.

Xu, P., Qiu, S., Cai, J., Li, C. Liu, H. 2017. A novel analytical solution for gas diffusion in multi-scale fuel cell porous media. *Journal of Power Sources* 362: 73–79.

Xu, P. Qiu, S., Yu, B.M., Jiang, Z. 2013. Prediction of relative permeability in unsaturated porous media with fractal approach. *International Journal of Heat and Mass Transfer* 64: 829–837.

Xu, P., Sasmito, A.P., Yu, B.M., Mujumdar, A.S. 2016. Transport phenomena and properties in treelike networks. *Applied Mechanics Reviews* 68: 040802.

Xu, P., Yu, B.M. 2008. Developing a new form of permeability and Kozeny-Carman constant for homogeneous porous media by means of fractal geometry. *Advances in Water Resources* 31: 74–81.

Xu, P., Yu, B.M., Qiao, X., Qiu, S., Jiang Z. 2013. Radial permeability of fractured porous media by Monte Carlo simulations. *International Journal of Heat and Mass Transfer* 57: 369–374.

Yano, T., and Nagai, T. 1989. Fractal surface of starchy materials transformed with hydrophilic alcohols. *Journal of Food Engineering* 10: 123–133.

Yiotis, A. G., Tsimpanogiannis, I. N. and Stubos, A. K. 2010. Fractal characteristics and scaling of the drying front in porous media: a pore network study. *Drying Technology* 28: 981–990.

Yu, B. M., and Cheng P. 2002a. A fractal model for permeability of bi-dispersed porous media. *International Journal of Heat and Mass Transfer* 45: 2983–2993.

Yu. B.M., and Cheng, P. 2002b. Fractal models for the effective thermal conductivity of bidispersed porous media. *Journal of Thermophysics and Heat Transfer* 16: 22–29.

Yu, B.M., and Li, J. 2004. A fractal model for the transverse thermal dispersion conductivity in porous media. *Chinese Physics Letters* 21: 117–120.

Yu, B.M., Li, J., Li, Z., Zou, M. 2003. Permeabilities of unsaturated fractal porous media. *International Journal of Multiphase Flow* 29: 1625–1642.

Yu, B.M., Xu, P., Zou M., Cai J. Zheng Q. 2014. *Fractal Physical Transport in Porous Media*. Beijing: Science Press.

Yu, B.M., Zou M., Feng, Y. 2005. Permeability of fractal porous media by Monte Carlo simulations. *International Journal of Heat and Mass Transfer* 48: 2787–2794.

Yuan Y., Zhang J., Wang D., Xu, Y., Bhandari, B. 2018. Molecular dynamics simulation on moisture diffusion process for drying of porous media in nanopores. *International Journal of Heat and Mass Transfer* 121: 555–564.

Zhang, N., and Wang, Z. 2017. Review of soil thermal conductivity and predictive models. *International Journal of Thermal Sciences* 117: 172–183.

Zhu, W.X., Dong, T.Y., Cao, C.W., Li, D. 2004. Fractal modeling and simulation of the developing process of stress cracks in corn kernel. *Drying Technology* 22: 59–69.

4 Heat and Mass Transfer Simulation of Intensification Drying Technologies: Current Status and Future Trends

Xian-Long Yu
China Agricultural University, Beijing, China

Hong-Wei Xiao
China Agricultural University, Beijing, China

Arun S. Mujumdar
McGill University & Western University,
Montreal, Quebec, Canada

CONTENTS

4.1   INTRODUCTION: BACKGROUND AND DRIVING FORCES

Drying is one of the oldest and most widespread processing techniques for products including food, textile, paper, wood, ceramics, minerals, wastewater sludge, pharmaceutical or biotechnological products and products that are considered 'highly perishable' (Mujumdar, 2006). In order to reduce energy consumption and improve food safety and security, intensification of drying is necessary. Although numerous options exist to intensify heat and mass transfer, in this chapter we will focus on microwave drying, air-impingement drying, radio frequency drying, superheat steam drying and infrared drying.

Drying is a complex multi-physics phenomena involving multi-scale and multiphase. The heat and mass transfer characteristics of the drying process are extremely complicated and diversified. There is now increased interest in exploring the heat and mass transfer characteristics in intensified drying. The main task of enhancing heat and mass transfer is to speed up the dehydration rate, which implies smaller equipment for a given production rate.

Numerical simulation involves solution of the governing mathematical description of the specific drying operation to obtain the change of physicochemical characteristics such as the moisture diffusion, heat transfer and material shrinkage. That can help us understand the foundations of drying theory and identify the information required to design a successful drying system. However, the numerical simulation cannot supplant experimental verification in view of our lack of detailed knowledge of the fundamental drying principles. Moreover, the capabilities of the user's computer also limit the application of numerical simulation. As a result, the development of numerical simulation has gone through a long process.

This chapter summarizes the research status of numerical simulation of intensification drying technologies as in Figure 4.1. It reviews the research of numerical models in different intensification drying technologies and discusses the strategies

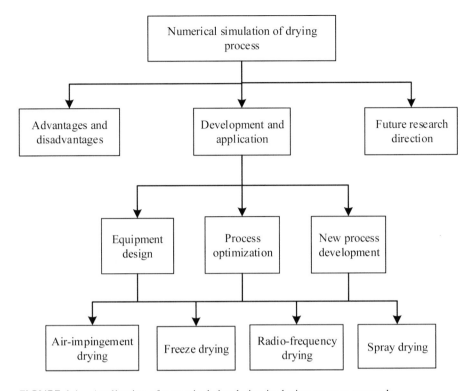

FIGURE 4.1 Application of numerical simulation in drying process research.

for enhancing performance based on the main findings. Some examples of model-based innovation are given along with brief tabulated summaries of various modeling approaches and a list of recent literature reports on modeling dryers that have appeared in *Drying Technology*. Finally, the challenges and research opportunities are identified and discussed.

4.2   NUMERICAL SIMULATION IS A POWERFUL TOOL

Drying is a process controlled by multiple factors. There are various factors affecting the heat and mass transfer in different drying methods. For convection, the temperature, humidity and flow velocity of the drying medium are the main factors affecting drying; in addition to the above factors, we need to consider the intensity of radiation and radiation distance for radiation drying; for spray drying, the particle size and the motion are important factors influencing drying rate and drying uniformity. The environment pressure, load volume, dryer structure and physicochemical properties of materials play an important role in drying. Through complex experimental design, we perhaps find a suitable drying process or mechanical structure for given specific drying method. However, it is difficult to determine the mechanism of the drying process and provide a basis for the drying of other materials. Numerical simulation can be used to explore phenomena and laws that are difficult to observe in the process of experiment and offset the shortcomings of experimentation.

The list of recent literature reports on modeling dryers that have appeared in *Drying Technology* is shown in Table 4.1. Numerical simulation can be applied to various drying techniques, like spray drying, superheated steam drying, convective drying, solar drying, infrared drying, microwave drying and freeze drying. The materials involved in the numerical models are also widely distributed, including fruits and vegetables, meat, dairy products, grain and paper. In addition, the numerical simulation can be used to explore a wide range of problems.

Numerical simulation can accurately describe the heat and mass transfer of the drying process in a different space and time. In comparison to experiments limited by the sensor technology, numerical simulation is considered an effective and useful tool for understanding heat and mass transfer mechanisms. Firstly, numerical simulation can obtain accurate and comprehensive results, which are quantified throughout the space and time, especially in the sharp variation zone. Then, the distribution of all variables (temperature, moisture content, velocity etc.) within the product and its surroundings is available at very high spatial and temporal resolution, even at scales or locations that are difficult to access experimentally. For example, through numerical simulation, the distribution of moisture and temperature in the material can be obtained at each moment during the whole drying process. The migration law of moisture and energy in dry materials can be clearly revealed by comparing the values at different time and space points.

The drying time, product qualities and energy consumption are important indexes used to evaluate drying technology. In order to obtain the appropriate processing parameters corresponding to the drying materials, it is usually necessary to optimize the drying conditions. It is difficult to obtain the optimal drying process through experimental methods under the influence of experimental cycle, cost and material properties diversity. On the contrary, numerical simulation has several advantages in this area of study, such as low cost, no-external interference and convenient data collection. Moreover, numerical simulating analysis does not require extensive experimental operations, but only a credible mathematical model. Modeling does not require additional lab space or maintenance and technical personnel to operate the equipment, and it avoids the high energy costs inherent to drying. In addition, the use of multi-objective process optimization methods can turn the numerical model into a proactive design and analysis tool.

Numerical simulations have no scale-up problems. As the processes occurring are highly nonlinear, the experimental results in laboratory or pilot plant setups are often less referential for practical application. Numerical simulations can be performed at the actual scale. It is possible to examine design tradeoffs and suggest modifications to improve the efficiency of drying and quality of the final products without the cost and effort of constructing a scaled or full-scale setup and the time needed to perform experiments.

Although numerical simulations show many unique advantages in heat and mass transfer of the drying process, the application of numerical simulation in drying research also has many limitations. Numerical modeling of dehydration of porous materials requires knowledge of material properties. Due to a large number of

TABLE 4.1
Recent Literature Reports on Modeling Dryers in *Drying Technology*

Author(s)	Drying Method	Material	Purpose	Dimension	Software
Jubaer et al. (2018)	spray drying	milk	droplet shrinkage behavior	3D	ANSYS Fluent
Le et al. (2018)	superheated steam drying	wood particles	establish simple drying model	2D	Academic
Fakhfakh et al. (2018)	convective drying	bovine leather	explore elastic mechanical behavior	2D	COMSOL
Sanghi et al. (2018)	solar drying	corn	explore the temperature, humidity and air velocity profiles	3D	ANSYS Fluent
Da Silva et al. (2017)	convective drying	ripe banana	describe continuous and intermittent	1D	Academic
Scaar et al. (2016)	mixed-flow drying	wheat	the effect of bed materials and air duct arrangements	3D	ANSYS CFX
Amantea et al. (2018)	convective drying	grains	energy and energy analyze	2D	Academic
Wu et al. (2017)	infrared drying	rice kernel	predict temperature, MC and internal stresses of rice kernels	3D	COMSOL
Aktas et al. (2017)	infrared drying	apricot	reducing energy demand	3D	ANSYS Fluent
Khomwachirakul et al. (2016)	impinging stream drying	particle	predict the particle motion behavior	3D	ANSYS Fluent
Tsuruta et al. (2015)	microwave-vacuum drying	radishes, carrots, tofu	the mechanism of deformation due to shrinkage of the food structure	2D	Academic
Jin et al. (2015)	microwave-vacuum drying	fresh stem lettuce	improve the uniformity of particle mixing and microwave heating	2D	ANSYS Fluent
Cumhur et al. (2016)	freeze drying	turkey breast meat	describe quantitatively the dynamic behavior of the primary and secondary drying stages	2D	MATLAB
Dolati et al. (2018)	electrohydrodynamic drying	porous body	study the flow patterns and the moisture content	2D	Academic
Keshani et al. (2015)	spray drying	none	investigate the effect of spray dryer geometry	3D	ANSYS Fluent
Rezk et al. (2015)	vacuum drying	paper sheet	characterize flow resistance in paper sheet and wire.	3D	COMSOL

influencing factors and computer performance limitations, it is necessary to simulate the drying process by taking all factors into account. However, the hypothesized conditions adopted in the simulation process reduce the accuracy of the results. To ensure the accuracy of the simulation, the numerical simulation usually needs the verification of experimental data, which limits the application of numerical simulation in the drying field. Numerical simulation is a powerful tool. To facilitate ease application and accuracy, researchers continue to refine numerical models and use them to enhance the performance of drying.

4.3 DEVELOPMENT AND APPLICATION OF NUMERICAL SIMULATION IN INTENSIFICATION DRYING TECHNOLOGIES

The mathematical model should be established for the numerical simulation of the drying process, which varies with the selected drying technology and material properties, as well as the objective question for researches. Table 4.2 provides the common models in the numerical simulation of drying, including the turbulence model for fluid flow, the heat transfer model for energy transfer, the mass transfer model for material migration and the shrinkage model for describing the shrinkage of materials in the drying process. Using the appropriate model will result in a high degree of agreement with the actual drying. The parameters involved in the model need to be obtained through experiments or literature review.

The numerical simulation has been further developed with the development of drying theory. Numerical models no longer focus on simple macroscopic phenomena, and vital forces are given by the rapid development of the micro-scale model and the multi-scale model. In the following section the development of numerical simulations of heat and mass transfer and the enhancement of several existing drying technologies is discussed.

4.3.1 Air-impingement Drying

Jet impingement systems provide an effective means to enhance the convective processes, which has been used when drying textiles, paper, food; blanching fruits and vegetables; cooling gas turbine components and electronics; freezing tissue in cryosurgery and manufacturing (Xiao and Mujumdar, 2014). Compared with conventional convective drying, impingement has a higher heat transfer coefficient. In order to provide a larger jet area, multiple jets are usually used together. When the jets are spaced closely, the air flows from the jets highly interact with each other. The local transfer rate of the multiple impinging jets is not only affected by the jet configuration, orifice shape, inlet flow and jet-to-surface distance, but also geometric parameters such as the nozzle arrangement and jet-to-jet spacing. Even small modifications of the impinging jets setup might substantially change the flow characteristics and the heat transfer performance.

In order to enhance heat and mass transfer of air-impingement drying, it is necessary to find the flow-field distribution and drying characteristics under different geometric and physical parameters. Particularly, accurate prediction of turbulence development is absolute importance.

TABLE 4.2

Basic Heat and Mass Transfer Model of Drying

Model	Function	Sort	Describe
Turbulence model	The movement of medium or material	Spalart-Allmaras	Suitable to describe wall boundary layer flow
		SST k-ω	This model combines the standard k-ω model and k-ε model, and includes the cross flow dissipation integral and shear force
		RNG k–ε	The turbulence at low Reynolds number is calculated, and the rotational flow is taken into account
Heat transfer model	Describes heat transfer during drying	$q'' = k\dfrac{dT}{dx}$	Fourier law is used to solve the energy transport in medium due to temperature gradient
		$q'' = h\left(T_s - T_\infty\right)$	Newton's law of cooling is used to describe the transfer of energy due to the conduction and fluid flow
		$q'' = \varepsilon\sigma\left(T_s^4 - T_{sur}^4\right)$	Based on Stephen Boltzmann's law, the energy transfer through electromagnetic waves is described
Mass transfer model	Describes the transfer of material during drying	$N_A = h_m A_s\left(C_{A,s} - C_{A,\infty}\right)$	The model describe the transfer of mass in a fluid due to conduction and fluid motion
		$j_A = -\rho D_{AB}\nabla m_A$	Based on Fick's law of diffusion, material transfer due to concentration gradient is described
Shrinkage model	Material volume and shape change during drying	ideal shrinkage	The shrinkages of the materials are treated as a pure liquid
		stress-strain	The heat and moisture transport equations are coupled to the structural mechanics
		multi-phase transport	Neo-Hookean constitutive model is used in order to determine material deformation

4.3.1.1 Numerical Simulation Methods in Impinging Jets

At present, there are mainly three kinds of numerical simulation methods for the study of impinging jet: direct numerical simulation, shear stress transport simulation and large eddy simulation.

Direct numerical simulation (DNS) does not require modeling of turbulence, and it is a useful tool in the study of turbulence theory. Due to the huge calculations for complex turbulence, DNS is mainly used to simulated simpler two-dimensional slot jets at low and moderate Reynolds numbers. Hattori & Nagano (2004) performed DNS of impinging flows to solve the governing equations directly with heat transfer to investigate the flows' structures. Da Silva & Pereira (2008) analyzed the invariants of the velocity-gradient, rate-of-strain and rate-of-rotation tensors across the turbulent/nonturbulent interface by using a DNS of a turbulent plane jet at low Reynolds number, Re = 120.

The shear stress transport simulation (SST) shows quite accurate prediction results in the impingement jets. This model combines the standard k-ω model and k-ε model, and includes the cross flow dissipation integral and shear force. Chen et al. (2012) analyzed the flow field of turbulent impinging jet in a room by comparison of the RNG k–ε and SST k–ω model and found the SST k-ω model accords slightly better in the region close the impingement zone. By comparing the performance of various turbulent models, Zuckman & Lior (2005) also found SST models could produce better predictions of fluid properties in impinging jet flows. Up to now, the SST model has been used for analysis of flow field and heat transfer so as to optimize the structural design (Wen et al., 2016; Chougule et al., 2011; Končar et al., 2010).

Large eddy simulation (LES) is more appealing for simulations of three-dimensional turbulent impinging jets at higher Reynolds numbers. Its computational costs can be considerably reduced as compared to the DNS and the subgrid scale model can be established to calculate small scale vortices. With sufficient precision, LES may gain a better insight into flow, vertical and turbulence structure and their correlation with the local heat transfer in impinging flows. Hadžiabdić & Hanjalic (2008) developed a LES model for single round impinging jet and analyzed the specific dynamics of the thermal field and the wall heat transfer, and elucidated the formation and motion of vortices. Hällqvist (2006) applied LES without use of a subgrid scale model to analyze the effect of the shape of the mean inlet velocity profile and the wall heat transfer on convective vertical structures. The results showed that the model provided accurate results for simple geometries. Draksler et al. (2017) used LES simulations to investigate the instantaneous flow field, coherent structures and instantaneous heat transfer characteristics of multiple impinging jets in hexagonal configuration.

In addition, based on the above models, the prediction of water diffusion in impingement drying can be carried out with an evaporation kinetics equation. Based on the SST model, De Bonis & Ruocco (2007) numerically analyzed local heat and mass transfer in food slabs in air-impingement drying. The evaporation kinetics has been tackled by a simple Arrhenius notation.

4.3.1.2   The Numerical Simulation of Impingement Configuration to Intensify Drying

The jet pattern is one important factor in the design of impingement systems for the optimum heat transfer performance. De Bonis & Ruocco (2014) developed an axisymmetric transfer phenomena model to study the dependence of the surface heat and mass transfer on geometry arrangement and fluid dynamic regime. Caliskan et al. (2014) investigated the velocity distributions over a smooth surface in six configurations of jet plates with different aspect ratios by simulation. It was found that the turbulent kinetic energy distributions are strongly influenced by the nozzle geometry and circular nozzles have better turbulent kinetic energy than rectangular nozzles.

The heat transfer characteristics greatly depend on the jet-to-jet spacing (*L/D*) of multiple jets. Olsson et al. (2005) simulated flow around two and three jets with a *k*-ε SST turbulence model and the result showed that the heat transfer coefficient is dependent on the flow behavior. The highest average value was found for cylinders under two impinging jets with a mutual distance of *L/D* = 2. For longer and shorter distances, the heat transfer is reduced.

4.3.1.3   The Numerical Simulation of the Airflow Pattern to Intensify Drying

Jet impingement provides an effective way to strengthen heat and mass transfer due to its thin hydrodynamic and thermal boundary layers in the stagnation region. However, the thin boundary layer is still a limit to the efficiency of a convective heat transfer process. Unsteady flow pattern can break boundary layer instabilities in jet impingement.

In the earlier experimental studies, some studies have found that pulsating jets can increase the drying rate, while others have found the opposite. To explain the cause of this phenomenon, the turbulence model is used to explore the development of airflow and results show that jet Reynolds numbers, frequency and time ratio of the intermittent pulsation, waveform of the impinge jets, jet temperature and geometrical configuration all affect the heat and mass transfer.

In general, the ability of convective heat transfer is assessed by a local Nusselt number and a local time-averaged Nusselt number. The simulation results show that the pulsating flow periodically alters the flow patterns and eliminates the formation of a static stagnation point and enhances the local Nusselt number, which perhaps enhances the degree of transient convection heat transfer.

In order to further enhance the heat and mass, the *k*-ω model has received attention in modeling turbulent impingement flows. Numerical simulation had been performed on a two-dimensional sinusoidal pulsating turbulent impinging jet under different temperatures, jet Reynolds numbers, amplitudes and frequencies of pulsation and nozzle-to-plate (Xu et al., 2010a). The single sinusoidal pulsation has no obvious difference from the steady turbulent impinging jet in time-averaged Nusselt numbers in the current situation, except for cases with large temperature differences. The sinusoidal pulsating turbulent impinging jet has been shown to display no noticeable heat transfer enhancement. Compared with sinusoidal pulsating jet, intermittent pulsating impinging jet shows stronger flow instabilities. The main factors affecting local time-averaged Nusset numbers was studied by means of numerical simulation (Xu et al., 2010b). The numerical results showed significant enhancement of heat transfer at the target surface by the intermittent pulsation in a turbulent impinging jet over a wide range of parameters for both cooling and heating cases.

In order to explore the potential effect of unsteady intermittent pulsation on the heat and mass transfer of multiple impinging jets, the RNG *k*-ε turbulence model takes account of turbulent vortices, and has been proven to be effective in modeling the type of complex flows in impinging and opposed jets. The numerical simulation was used to calculate the velocity field, which performed on a two-dimensional pulsations jets array associated with intermittent pulsations (Xu et al., 2012). The numerical results showed that the pulsations with phase angle difference could significantly improve the convective heat transfer rate. The intermittent pulsation frequency has significant effect on the local heat transfer, which should be in the range of the vortex frequency of the turbulence.

4.3.2 FREEZE DRYING

Freeze drying minimizes losses of flavor and nutritional composition and keeps the organoleptic properties of the initial fresh products due to the absence of liquid water and to the low temperature used in the dehydration process, which hinders most of the deterioration and microbiological reactions. The most notable advantages of freeze drying are the preservation of the intrinsic micro-scale structure or the formation of high porosity, minimization of thermal and chemical degradation, retention of volatile or aromatic components etc. Since the original physical structure of the material is retained, it can be reconstituted easily and rapidly by the addition of water. Thus, freeze drying is generally accepted as the best method to dehydrate high-quality and heat-sensitive products such as quality foods, pharmaceuticals and biomedical products.

Freeze drying utilizes sublimation of ice as its main drying mechanism. The process consists of freezing the raw material and removing the moisture by sublimation and desorption. The required energy for the sublimation is provided by conduction or radiation such as electric heating, infrared or microwave. In the drying process, the moisture in products is frozen in the pre-freezing stage and the ice is removed by sublimation at low temperature and low pressure during the 'primary drying' stage of the process. The sublimation interface moves through the material until only a dried material remains at the end of primary drying. Finally, water vapor diffuses through the pores to exit the drying material.

Despite its merit with respect to quality, freeze drying is the most sophisticated and expensive process of all drying techniques because of the slow drying rates and the high energy consumption. The drying process times are often quite long. Although a higher ice temperature is good for the drying rate, excessive ice temperatures may lead to severe loss of product quality and rejection of the batch. Therefore, reducing freeze-drying time in order to reduce energy consumption and raise productivity has been a worldwide challenge during the past decades. A precise knowledge of the fluid dynamic, heat and mass transfer mechanism in vacuum freeze drying is important to adjust the process variables to enhance heat and mass transfer.

4.3.2.1 The Development of a Numerical Simulation of Freeze Drying

The first freeze-drying model was the uniformly retreating ice front (URIF) model established by King (1971). The model can explain the drying rate when removing free water from materials and is intended to describe drying rates during the removal of the first 75–90% of the moisture. However, the URIF model has been shown to be quite inaccurate when it was used to predict drying times, as the process of removing adsorbed water and crystal water was ignored.

Sheng & Peck (1975) developed a freeze-drying model that accounts for the time required to remove the ice as well as all the bound water in products. The bound water is assumed to be a linear function of the temperature at all points in the ice-free region. However, this model is limited by the assumptions (1) of constant surface and interface temperature, (2) of a heat transfer–controlled process and (3) that the combined water is vaporized after sublimation of all free water.

Litchfield & Liapis (1979) constructed an adsorption-sublimation model using material and energy balances in the dried and frozen regions of a material undergoing freeze drying. Both sublimation and adsorption were accounted for in the set of coupled nonlinear partial differential equations. The mathematical models allowed desorption and sublimation to take place simultaneously, so that an absorbed moisture gradient occurs across the dried layer as drying proceeds. The adsorption-sublimation model was seen to be more accurate than the URIF model, and it is possible to predict the drying time quite accurately.

The fixed grid methods can provide a simpler approach to freeze-drying problems; however, the accuracy of results is poor. Difficulties arise in the treatment of the evolution of the complex sublimation interface during the freeze drying of slab-shaped products. Moving grid methods based on the explicit tracking of the interface position and exact imposition of the interfacial condition are believed to be the most suitable for dealing with the freeze-drying problems, e.g., freeze drying of solutions in vials, which were successfully solved by the moving grid methods (Sheehan & Liapis, 1998; Nam & Song, 2007). The sublimation interface moves through the material until only a dried material remains at the end of primary drying. At the end of the primary drying stage, all ice has been removed.

4.3.2.2 The Numerical Simulation of Technological Parameters to Intensify Drying

There are a few methods to enhance heat and mass transfer while maintaining an acceptable quality of freeze-dried products, such as optimum heating temperature, control desorption process and multiple heating combinations.

The sorption-sublimation model is the most commonly used mathematical model to simulate the heat and mass transfer process of freeze drying. The model can be used to simulate the freeze-drying process under various operational conditions to reduce the drying times. By using the model, the temperature and absorbed water data can be predicted accurately to determine operational conditions that may minimize not only the drying time but also the deteriorative changes. In addition, the criterion used in terminating the freeze-drying process is of extreme importance, since it may lead to an undesirable absorbed water profile which may deteriorate the quality of the dried product.

The thin sheet material is favorable for reducing both the drying time and the sublimation temperature by decreasing the diffusion length for vapor transport and increasing the interfacial area for sublimation. Nam & Song (2007) numerically studied the freeze-drying characteristics of planar and slab-shaped food products using a simulation program that considered the conjugate heat and mass transfer, sublimation of ice and motion of sublimation interface. The results showed that the lateral permeable surface of the slab-shaped products significantly altered the freeze-drying characteristics by reducing the diffusion length for the water vapor transfer as well as by increasing the interfacial area for sublimation.

The combined use of various heating methods, such as infrared heating, is a major way to enhance heat transfer in freeze drying. The mathematic model, based on the radiant heat transfer formula, can be used to analyze the heat and mass transfer of the freeze-drying processes with infrared radiation heating. Bae et al. (2010) proposed a spectral two-radiation model based on the fixed grid numerical framework. The model successfully solved steady and transient conduction radiation problems and validated the model for radiative transfer and energy conservation equations. The study also discussed the effect of heater temperature on freeze-drying behaviors. The numerical model is expected to be a useful tool for predicting and optimizing freeze-drying processes with infrared radiation heating.

4.3.3 RADIO FREQUENCY DRYING

Radio frequency drying involves utilizing electromagnetic energy at a frequency range of 3 kHz to 300 MHz, which is based on the interaction between the electric field produced by the electrodes of a capacitor and the dipoles and ionic charges contained within the product. Radio frequencies (RF) can penetrate into the material, causing the migration of charged ions in the material, and then achieve heating by converting electrical energy into heat energy. Radio frequency heating offers several advantages such as rapid heating, uniform heating in whole material, efficient energy conversion and rapid process control. At present, radio frequency drying has been used in commercial applications in drying operations for various types of non-food materials, including textiles, wood and paper. Industrial applications can also be found in drying operations for food products, e.g., almonds, dates, lentils, raisins, spices and wheat.

Because of their long wavelengths, radio frequencies can achieve deeper heating than microwaves, but the strong edge effect is still a problem for radio frequency heating technology to be employed in the food industry. In the radio frequency drying process, the dielectric loss factor of the material gradually increases with the increase of the temperature, and the absorbed radio frequency energy increases with the increase of dielectric loss factor. Therefore, the energy is concentrated in the part with higher local temperature, which may lead to significant non-uniform heating. If the shape of the material is irregular, the radio frequency energy will be concentrated in the part with the greatest thickness in the material. Therefore, it is critical to design applicators that provide uniform RF field patterns in foods.

Radio frequency drying is typically an unsteady and irreversible process. The mass and heat transfer mechanisms are more difficult and complex. The numerical simulation is beneficial for assisting with the design and describing the changes of main substances during the drying process.

4.3.3.1 Numerical Simulation Development of Radio Frequency Heating

The radio frequency drying model is multi-physics, which involves the electromagnetic equation and heat transfer equation. The electrical field is a key to model the radio frequency drying process. Neophytou & Metaxas (1996) and Sheehan & Liapis (1998) established the electrical field model for industrial-scale RF heating systems by solving the coupled Laplace and wave equations. The simulation model makes it possible to obtain heat transfer from various agricultural products.

A challenge in using numerical simulation to analyze radio frequency systems is to validate the simulated results with experiments. Chan et al. (2004) studied numerical simulation of electromagnetic fields during radio frequency heating using the wave equation. It was found that the numerical data agreed well with the experimental data on the phase, heating patterns for different sized loads and positions.

The heat and mass transfer simulation of radio frequency heat treatment can be realized by solving the coupled diffusion and electromagnetic equations. Yang et al. (2003) solved coupled electromagnetic and diffusion equations by using the TLM-FOOD-HEATING program to simulate the temperature distribution of seeds by radio frequency energy. Marra et al. (2007) presented and validated a model of radio frequency heating of a foodstuff by FEMLAB. Many commercial packages are available for simulating radio frequency heating, such as HFSS from Ansoft (Pitts, PA, USA), COMSOL Multiphysics from Comsol (Los Angeles, CA, USA) and ANSYS from Ansys Corp. (Canonsburg, PA, USA). For example, COMSOL Multiphysics software, based on the finite element method, offers a powerful solver by enabling the simulation of coupled phenomena, which can be used to achieve the numerical simulation of wood heating process in radio frequency field (Laza Bulc et al., 2014).

4.3.3.2 Application of Numerical Simulation to Enhance Uniformity Radio Frequency Heating

The non-uniform heating in radio frequency treated products may cause incomplete or excessive drying. In order to explore the internal temperature and moisture content of materials, numerical simulation has been conducted to study the uniformity and related factors of radio frequency drying in various drying materials. Non-homogeneous heating was detected inside the samples and different behaviors were observed for different parameters by numerical simulation. Several models have been developed to improve the radio frequency uniformity for different materials.

The correlations of the material characteristics and electromagnetic field characteristics derived from a simulation model can provide valuable information and strategies to improve the radio frequency heating uniformity. Jiao et al. (2015) developed a simulation model for radio frequency heating of wheat kernels to study the influence of sample size and identified cold and hot spot positions affected by sample vertical position. Tiwari et al. (2011) investigated the influence of various factors on radio frequency power distribution in dry food materials, including sample size, shape, relative position and dielectric properties, using a validated finite element computer model. The study results showed that the uniformity of radio frequency heating is closely related to sample size, shape, dielectric coefficient and vertical position.

The material parameters have great influence on the uniformity of radio frequency heating. A large, cuboid-shaped sample i placed on the bottom electrode has better temperature uniformity than that of a small sized sample. For samples with small size, samples close to the electrode would achieve better radio frequency heating uniformity. The sample shape also affected the radio frequency power distribution and uniformity. A cuboid sample had higher RF power densities at the edges, while an ellipsoid had higher power densities in the center parts. The smaller values of dielectric properties resulted in better radio frequency power uniformities in the samples. Proper electrode parameters can also improve the uniformity. Reducing the electrode gap as well as a particular top electrode bending position and angle can improve the radio frequency power uniformity of the sample. The mathematical models for radio frequency heating can be used to demonstrate better container material and design to ensure good temperature distribution in practical radio frequency treatments. Huang et al. (2016) used computational modeling to attempt to quantify the effects of dielectric properties and densities surrounding a container in a parallel plate radio frequency system to improve the heating uniformity of dry food products. The optimum radio frequency heating uniformity in products could be achieved with a smaller density value of the surrounding container, which has smaller loss factor values.

4.3.3.3   Numerical Simulation of the Combination of Radio Frequency and Conventional Drying

The combination of radio frequency and conventional drying maximizes strengths and minimizes weaknesses of each method used alone. Compared with radio frequency drying, the combination drying provides a higher dehydration rate and more consistent heat and mass transfer. The radio frequency provides the heat to facilitate moisture transfer from inside of the products to the surface, where airflow is used to remove moisture from the surface.

Marshall & Metaxas (1999) developed and validated a boundary value approach to model the radio frequency electric field strength during the radio frequency-assisted heat pump drying of particulate materials. The results show that the combination system improves the dryer and heat pump performance. The ability of the radio frequency to source the internal moisture also improves the effectiveness of overall energy utilization.

Jumah (2005) established the heat and mass transfer model of radio frequency-assisted fluidized bed drying of grains, analyzing the change of temperature and moisture in material with time. Compared to pure convective drying, the radio frequency-assisted convective drying significantly increases the core temperature in the material, which increases the water diffusion rate from the interior to the surface. An increase in field frequency induces higher particle temperature and hence lower moisture content. As the material temperature cannot be raised indefinitely in the process temperature, the temperature distribution of materials can be predicted by numerical simulation, which can provide reference for the optimal setting of electric field intensity.

The temperature of air flow is an important factor affecting the heat and mass transfer rate. Through numerical simulation, the change of the internal temperature and moisture content of materials under different air flow temperatures over time can be compared. The effect of increasing the inlet air temperature is clearly to increase the particle temperature and hence the drying rate. To prevent condensation of the evaporated moisture on the surface of the particle, it is necessary to heat the fluidizing air to improve its moisture-carrying capacity.

Intermittent radio frequency heating is an efficient method to enhance moisture transfer from material surface to the drying medium. Numerical simulations can be used to demonstrate intermittent radio frequency heating with fluidizing air to find the change law of moisture content. In the non-radio frequency heating stage, the water diffusion rate on the material surface decreases rapidly, but the moisture content on the material surface increases gradually as the internal moisture of the material continues to diffuse to the surface. In the radio frequency heating stage, the water on the surface of the material quickly evaporates. Compared with continuous radio frequency heating, periodic radio frequency heating requires longer drying time. However, the intermittent radio frequency heating maintains low temperature in material that is more suitable for drying of thermally sensitive bio-products.

4.3.3.4   Numerical Simulation of Radio Frequency/Vacuum Drying

Radio frequency/vacuum drying produces a high-quality product with short drying time. The heat is generated directly within the products being dried and the vacuum environment creates a large pressure gradient that provides the drying force of mass transfer. The ambient pressure is very low in the radio frequency/vacuum drying process, and the internal temperature of the drying material increases rapidly under continuous radio frequency heating. When the temperature approaches and exceeds the boiling point, the liquid phase water begins to boil. The phase change rate of ebullition is much larger than that of evaporation.

According to the previous analysis of the heat and mass transfer mechanisms, the numerical simulation can serve as an efficient tool to quantitatively describe the transport phenomena of heat and mass during continuous radio frequency/vacuum drying, derived from conservation equations. Koumoutsakos et al. (2001) explored the moisture and heat transfer mechanism in wood during radio frequency/vacuum drying and established a one-dimensional mathematical model. The results indicated the internal moisture in wood was transferred by the driving force of pressure gradient in the form of vapor above fiber saturation point.

Jia et al. (2016) compared and analyzed the mass transfer law of radio frequency/vacuum drying under different pressures by numerical model. For a lower ambient pressure, most of the moisture is transferred along the longitudinal direction to the surroundings in the form of vapor, and the driving force due to differential pressure would slightly increase, successively leading higher drying rate. However, for a higher ambient pressure, the driving force has less influence on the drying rate and the relationship between the driving force and the drying rate becomes inexplicit.

The vacuum environment reduces the loss of radio frequency energy. Jia et al. (2017) provided a relatively fast and efficient model based on the mass and heat transfer theory of porous materials to simulate radio frequency/vacuum drying behavior, and presented the evolution curves of independent variables including moisture content, temperature, evaporation etc. In addition, for the initial drying stage, the temperature of each region in material rose rapidly due to the heating of radio frequency and the vapor mass flow rose rapidly with the raising temperature.

4.3.4 SPRAY DRYING

Spray drying is an important process in powder technology, which is widely used in the chemical industry, light industry and the food industry. These particles are subjected to drag forces, buoyancy etc. and also exchange heat and mass with their surroundings. However, the complexity of airflow patterns and particle migration inside the spray dryer make their physical design and process parameter selection difficult.

Numerical simulation has been used to investigate the performance and design of spray dryers particularly in the food industry, including the motion of particles, their residence time in the clamber, optimization of drying conditions and so on.

4.3.4.1 The Numerical Simulation of Equipment Design for Improving Droplets Movement

The motion state of particles in a spray dryer is an important factor affecting the dehydration rate and product quality. Numerical simulation can predict the trajectories of particles of different shapes and densities in specific structural environments. Oakley & Bahu (1991) predicted the trajectories of water droplets from a hollow-cone pressure nozzle in a chamber using numerical simulation. They noted the strong cooling effect of the evaporating droplets on the central gas jet, particularly under the atomizer where the central jet is partially sheltered from the hot-temperature inlet gas stream by the high droplet concentration around the atomizer. Langrish & Zbicinski (1994) demonstrated the use of a numerical model to explore methods for decreasing the wall deposition rate, including modifications to the air inlet velocity and a reduction in the spray cone angle. Southwell et al. (2001) have used numerical simulation to improve the flow distribution in a pilot spray dryer. This study investigated ways in which a uniform flow in the plenum chamber could be obtained for a configuration with a single, off-axis, inlet pipe. They used the CFX4 and CFX5 code to perform calculations for a variety of baffle configurations to determine the effect of various options without the need to carry out extensive experimental studies. These studies provide necessary theoretical references for improving the drying uniformity of the spray dryer.

In addition, designing a full-scale spray dryer involves serious difficulties and unreliability because of the complexity involved in measuring and describing the flow, heat and mass transfer mechanisms within the drying chamber. Dynamic similarity of airflow patterns is an important prerequisite to ensure that the performance of the machine does not change after scale up. The numerical simulation is also a powerful tool to assist with scale up for the spray dryer.

4.3.4.2 Application of Numerical Simulation to Enhance Heat and Mass Transfer in Spray Drying

In the spray-drying process, the drying medium transfers the heat to the particles, and the moisture in the particles moves to the drying medium. Knowledge of the spray-drying technology is essential for controlling its operational variables. The heat and mass transfer can be improved by improving spray dispersal, increasing the relative air-particle velocity and enhancing the exchange of air between the spray envelope and the surrounding hot and relatively dryer air. Harvie et al. (2001)

performed numerical simulations of the air flow patterns within a small scale tall-form countercurrent spray dryer, and found that by altering the angle of the inlet air streams into the dryer the nature of the flow within the dryer could be significantly altered. Southwell et al. (1999) investigated the effect of enhanced turbulence in the inlet air flow on overall drying performance using the CFD package CFX4 and found the trend of increased evaporative performance from increased inlet turbulence.

The type of drying medium is also an important factor influencing the heat and mass transfer rate during spray drying. Frydman et al. (1998) simulated a discrete droplet model of a pilot-scale spray dryer operated with two drying media: super-heated steam and air. In superheated steam, the droplets penetrated to a greater extent in the recirculation zone and evaporated faster.

The homogeneous morphology is an important index for evaluating product quality after spray drying. In order to control the morphology of granules, it becomes important to clarify their formation mechanisms. Nishiura et al. (2010) investigated the relationship between the morphology of spray-dried granules and drying conditions by using numerical simulation. The particle motion and gas-liquid flow during the drying of droplet-suspended particulates were calculated at various air temperatures and the results showed that a depression is formed at the granule surface at the part with low drying rate; in contrast, in the case of the part with the higher drying rate, the crust forms rapidly. To gain insight into the granule formation, Breinlinger et al. (2015) simulated the influence of surface tension on granule morphology during spray drying using a simple capillary force model. The results suggested that surface tension is an important parameter in spray drying and that it should not be neglected in product and process design.

4.4   FUTURE TRENDS AND CLOSING REMARKS

Numerical simulation has become an indispensable tool for drying equipment design and process optimization. However, there are still many aspects need to be improved.

Complex models are closer to reality but they are often more difficult to solve. In some cases, the simplest models fitted the results properly, which is why they are still valid. To enhance ease of use and accuracy, new tools or approaches need to be developed for optimizing the drying process and understanding the phenomenology of operation, such as the multi-scale models, the conjugated models and micro-modeling.

Drying is a complex phenomenon involving energy, mass and momentum transfer in porous media and heat and mass exchange between the material surface and the external environment. Numerical models should be used as a proactive design and analysis tool with their own advantages. In addition to exploring the heat and mass transfer mechanisms of existing drying technologies so as to strengthen drying processes, numerical simulation is expected to be used to explore more advanced drying technologies and processes.

At present, numerical simulation is mainly used in laboratory drying research, and many advanced drying techniques have not yet been applied to industry. However, the ultimate goal of drying research is to improve industrial production capacity. Therefore, the numerical simulation needs to be combined with industrial production to conduct in-depth research on the laws of heat and mass transfer under industrial production and to expand the laboratory research results to the industrial level. It is necessary to explore the numerical models and calculation methods suitable for the industrial level.

It is noted that the evaluation of drying effects requires comprehensive evaluation of drying rate, product quality, energy consumption and other indicators. When the structure design and process optimization of drying equipment are carried out using numerical simulation, the mathematical model should take into account the physical and chemical property change, energy destination and heat and mass transfer rules of materials comprehensively.

Numerical simulation is a powerful tool for drying equipment development and process optimization. At present, a great deal of research has been carried out on the heat and mass transfer mechanism of numerical simulation for various advanced drying technologies, and the numerical model has also made great progress. For better application in drying industry, more accurate mathematical models and more efficient solving methods are needed to improve the computational efficiency and scientificity of numerical simulation.

ACKNOWLEDGEMENT

This research was supported by the National Key Research and Development Program of China (No.2017YFD0400905).

REFERENCES

Aktas, M., Sozen, A., Amini, A., & Khanlari, A. (2017). Experimental analysis and CFD simulation of infrared apricot dryer with heat recovery. *Drying Technology*, 35: 766–783.

Amantea, R. P., Fortes, M., Ferreira, W. R., & Santos, G. T. (2018). Energy and exergy efficiencies as design criteria for grain dryers. *Drying Technology*, 36: 491–507.

Bae, S. H., Nam, J. H., Song, C. S., & Kim, C. J. (2010). A numerical model for freeze drying processes with infrared radiation heating. *Numerical Heat Transfer*, 58(5), 333–355.

Breinlinger, T., Hashibon, A., & Kraft, T. (2015). Simulation of the influence of surface tension on granule morphology during spray drying using a simple capillary force model. *Powder Technology*, 283, 1–8.

Caliskan, S., Baskaya, S., & Calisir, T. (2014). Experimental and numerical investigation of geometry effects on multiple impinging air jets. *International Journal of Heat & Mass Transfer*, 75(6), 685–703.

Chan, T. V., Tang, J., & Younce, F. (2004). 3-Dimensional numerical modeling of an industrial radio frequency heating system using finite elements. *Journal of Microwave Power and Electromagnetic Energy*, 39(2), 87–105.

Chen, H. J., Moshfegh, B., & Cehlin, M. (2012). Numerical investigation of the flow behavior of an isothermal impinging jet in a room. *Building & Environment*, 49(3), 154–166.

Chougule, N. K., Parishwad, G. V., Gore, P. R., & Pagnis, S. (2011). CFD analysis of multi-jet air impingement on flat plate. *Lecture Notes in Engineering & Computer Science*, 2192(1), 2431–2435.

Cumhur, O., Seker, M., & Sadikoglu, H. (2016). Freeze drying of turkey breast meat: Mathematical modeling and estimation of transport parameters. *Drying Technology*, 34: 584–594.

da Silva, C., & Pereira, J. C. F. (2008). Invariants of the velocity-gradient, rate-of-strain, and rate-of-rotation tensors across the turbulent/nonturbulent interface in jets. *Physics of Fluids*, 20(5), 765.

da Silva, W. P., Rodrigues, A. F., e Silva, C. M. D. P. S., & Gomes, J. P. (2017). Numerical approach to describe continuous and intermittent drying including the tempering period: Kinetics and spatial distribution of moisture. *Drying Technology*, 35: 272–280.

De Bonis, M. V., & Ruocco, G. (2007). Modelling local heat and mass transfer in food slabs due to air jet impingement. *Journal of Food Engineering*, 78: 230–237.

De Bonis, M. V., & Ruocco, G. (2014). Conjugate heat and mass transfer by jet impingement over a moist protrusion. *International Journal of Heat and Mass Transfer*, 70: 192–201.

Draksler, M., Končar, B., Cizelj, L., & Ničeno, B. (2017). Large eddy simulation of multiple impinging jets in hexagonal configuration – flow dynamics and heat transfer characteristics. *International Journal of Heat & Mass Transfer*, 109: 16–27.

Dolati, F., Amanifard, N., & Mohaddes Deylami, H. (2018). Numerical investigation of mass transfer enhancement through a porous body affected by electric field. *Drying Technology*, 36: 1563–1577.

Fakhfakh, R., Mihoubi, D., & Kechaou, N. (2018). Numerical modeling assessment of mechanical effect in bovine leather drying process. *Drying Technology*, 36: 1313–1325.

Frydman, A., Vasseur, J., Moureh, J., Sionneau, M., & Tharrault, P. (1998). Comparison of superheated steam and air operated spray dryers using computational fluid dynamics. *Drying Technology*, 16(7), 1305–1338.

Hadžiabdić, M., & Hanjali, K. (2008). Vortical structures and heat transfer in a round impinging jet. *Journal of Fluid Mechanics*, 596(596), 221–260.

Hällqvist, T. (2006). Large eddy simulation of impinging jets with heat transfer. PhD Thesis, Royal Institute of Technology, Department of Mechanics, Sweden.

Harvie, D. J. E., Langrish, T. A. G., & Fletcher, D. F. (2001). Numerical simulations of gas flow patterns within a tall-form spray dryer. *Chemical Engineering Research & Design*, 79(3), 235–248.

Hattori, H., & Nagano, Y. (2004). Direct numerical simulation of turbulent heat transfer in plane impinging jet. *International Journal of Heat & Fluid Flow*, 25(5), 749–758.

Huang, Z., Marra, F., & Wang, S.J. (2016). A novel strategy for improving radio frequency heating uniformity of dry food products using computational modeling. *Innovative Food Science & Emerging Technologies*, 34: 100–111.

Jia, X., Hayashi, K., Zhan, J., & Cai, Y. (2016). The moisture transfer mechanism and influencing factors in wood during radio-frequency/vacuum drying. *European Journal of Wood & Wood Products*, 74(2), 1–8.

Jia, X., Zhao, J., & Cai, Y. (2017). Mass and heat transfer mechanism in wood during radio frequency/vacuum drying and numerical analysis. *Journal of Forestry Research*, 28(1), 1–9.

Jiao, S., Deng, Y., Zhong, Y., Wang, D., & Zhao, Y. (2015). Investigation of radio frequency heating uniformity of wheat kernels by using the developed computer simulation model. *Food Research International*, 71: 41–49.

Jin, G. Y., Zhang, M., Fang, Z. X., Cui, Z. W., & Song, C. F. (2015). Numerical Investigation on Effect of Food Particle Mass on Spout Elevation of a Gas-Particle Spout Fluidized Bed in a Microwave-Vacuum Dryer. *Drying Technology*, 33: 591–604.

Jubaer, H., Afshar, S., Xiao, J., Chen, X. D., Selomulya, C., & Woo, M. W. (2018). On the importance of droplet shrinkage in CFD-modeling of spray drying. *Drying Technology*, 36: 1785–1801.

Jumah, R. (2005). Modelling and simulation of continuous and intermittent radio frequency-assisted fluidized bed drying of grains. *Food & Bioproducts Processing*, 83(3), 203–210.

Keshani, S., Montazeri, M. H., Daud, W. R. W., & Nourouzi, M. M. (2015). CFD Modeling of Air Flow on Wall Deposition in Different Spray Dryer Geometries. *Drying Technology*, 33: 784–795.

Khomwachirakul, P., Devahastin, S., S Swasdisevi, T., & Soponronnarit, S. (2016). Simulation of flow and drying characteristics of high-moisture particles in an impinging stream dryer via CFD-DEM. *Drying Technology*, 34: 403–419.

King, C. J. (1971). Freeze-drying of foods. CRC Press, Cleveland.

Končar, B., Norajitra, P., & Oblak, K. (2010). Effect of nozzle sizes on jet impingement heat transfer in he-cooled divertor. *Applied Thermal Engineering*, 30(6–7), 697–705.

Koumoutsakos, A., Avramidis, S., & Hatzikiriakos, S. G. (2001). Radio frequency vacuum drying of wood. i. mathematical model. *Drying Technology*, 19(1), 65–84.

Langrish, T. A. G., & Zbicinski, I. (1994). Effects of air inlet geometry and spray cone angle on the wall deposition rate in spray dryers. *Chemical Engineering Research & Design*, 72(A3), 420–430.

Laza Bulc, M., Leuca, T., & Slovac, F. (2014). Numerical modeling of the heating process in a radiofrequency electromagnetic field using comsol multiphysics. *Journal of Electrical & Electronics Engineering*, 7(2), 17–20.

Le, K. H., Hampeln, N., Kharaghania, A., Buck, A., & Tsotsas, E. (2018). Superheated steam drying of single wood particles: A characteristic drying curve model deduced from continuum model simulations and assessed by experiments. *Drying Technology*, 36: 1866–1881.

Litchfield, R. J., & Liapis, A. I. (1979). An adsorption-sublimation model for a freeze dryer. *Chemical Engineering Science*, 34(9), 1085–1090.

Marra, F., Lyng, J., Romano, V., & Mckenna, B. (2007). Radio-frequency heating of food-stuff: solution and validation of a mathematical model. *Journal of Food Engineering*, 79(3), 998–1006.

Marshall, M. G., & Metaxas, A. C. (1999). A novel radio frequency assisted heat pump dryer. *Drying Technology*, 17(7–8), 1571–1578.

Mujumdar, A.S. (2006). Impingement drying. In Mujumdar, A.S. (Ed.), *Handbook of Industrial Drying* (third edition). pp. 385–394. CRC Press, Boca Raton, USA.

Nam, J. H., & Song, C. S. (2007). Numerical simulation of conjugate heat and mass transfer during multi-dimensional freeze drying of slab-shaped food products. *International Journal of Heat and Mass Transfer*, 50, 4891–4900.

Neophytou, R. I., & Metaxas, A. C. (1996). Computer simulation of a radio frequency industrial system. *Journal of Microwave Power*, 31(4), 251–259.

Nishiura, D., Shimosaka, A., Shirakawa, Y., & Hidaka, J. (2010). Simulation of drying of particulate suspensions in spray-drying granulation process. *Journal of Chemical Engineering of Japan*, 43(8), 641–649.

Norton, T., Sun, D.W. (2006). Computational fluid dynamics (CFD)–an effective, efficient design, analysis tool for the food industry: a review. *Trends of Food Science and Technology*, 17, 600–620.

Oakley, D. E., & Bahu, R. E. (1991). Spray/gas mixing behaviour within spray dryer. In Drying'91, Amsterdam, 1991, 303–313.

Olsson, E. E. M., Ahrné, L. M., & Trägårdh, A. C. (2005). Flow and heat transfer from multiple slot air jets impinging on circular cylinders. *Journal of Food Engineering*, 67(3), 273–280.

Rezk, K., Nilsson, L., Forsberg, J., & Berghel, J. (2015). Simulation of Water Removal in Paper Based on a 2D Level-Set Model Coupled with Volume Forces Representing Fluid Resistance in 3D Fiber Distribution. *Drying Technology*, 33: 605–615.

Sanghi, A., Ambrose, R. P. K., & Maier, D. (2018). CFD simulation of corn drying in a natural convection solar dryer. *Drying Technology*, 36: 859–870.

Scaar, H., Franke, G., Weigler, F., Delele, M., Tsotsas, E., & Mellmann, J. (2016). Experimental and numerical study of the airflow distribution in mixed-flow grain dryers. *Drying Technology*, 34: 595–607.

Sheng, T. R., & Peck, R. E. (1975). Rates for freeze-drying. *AIChE Symp. Ser.*, 73(163), 124–130.

Sheehan, P., & Liapis, A. I. (1998). Modeling of the primary and secondary drying stages of the freeze drying of pharmaceutical products in vials: Numerical results obtained from the solution of a dynamic and spatially multi-dimensional lyophilization model for different operational policies. *Biotechnology and Bioengineering*, 60, 712–728.

Southwell, D. B., Langrish, T. A. G., & Fletcher, D. F. (1999). Process intensification in spray dryers by turbulence enhancement. *Chemical Engineering Research & Design*, 77(3), 189–205.

Southwell, D. B., Langrish, T. A. G., & Fletcher, D. F. (2001). Use of computational fluid dynamics techniques to assess design alternatives for the plenum chamber of a small spray dryer. *Drying Technology*, 19(2), 257–268.

Tiwari, G., Wang, S., Tang, J., & Birla, S. L. (2011). Analysis of radio frequency (rf) power distribution in dry food materials. *Journal of Food Engineering*, 104(4), 548–556.

Tsuruta, T., Tanigawa, H., & Sashi, H. (2015). Study on Shrinkage Deformation of Food in Microwave-Vacuum Drying. *Drying Technology*, 33: 1830–1836.

Wen, Z. X., He, Y. L., Cao, X. W., & Yan, C. (2016). Numerical study of impinging jets heat transfer with different nozzle geometries and arrangements for a ground fast cooling simulation device. *International Journal of Heat & Mass Transfer*, 95, 321–335.

Wu, J. Z., Zhang, H. W., & Li, F. (2017). A study on drying models and internal stresses of the rice kernel during infrared drying. *Drying Technology*, 35: 680–688.

Xiao, H.W., & Mujumdar, A.S. (2014). Chapter 12: Impingement drying: applications and future trends. In *Drying Technologies for Foods: Fundamentals & Applications*. Prabhat K. Nema, Barjiinder Pal Kaur, & Arun S. Mujumdar (Eds.). pp. 279–299. New India Publishing Agency, New Delhi, India.

Xu, P., Mujumdar, A., Poh, H., & Yu, B. (2010a). Heat transfer under a pulsed slot turbulent impinging jet at large temperature differences. *Thermal Science*, 14(1), 271–281.

Xu, P., Qiu, S. X., Yu, M. Z., Qiao, X. W., & Mujumdar, A. S. (2012). A study on the heat and mass transfer properties of multiple pulsating impinging jets. *International Communications in Heat and Mass Transfer*, 39, 378–382.

Xu, P., Yu, B., Qiu, S., Poh, H. J., & Mujumdar, A. S. (2010b). Turbulent impinging jet heat transfer enhancement due to intermittent pulsation. *International Journal of Thermal Sciences*, 49(7), 1247–1252.

Yang, J., Zhao, Y., & Wells, J. H. (2003). Computer simulation of capacitive radio frequency (rf) dielectric heating on vegetable sprout seeds. *Journal of Food Process Engineering*, 26(3), 239–263.

Zuckerman, N., & Lior, N. (2005). Impingement heat transfer: correlations and numerical modeling. *Journal of Heat Transfer*, 127(5), 544–552.

5 Thermal Drying Utilizing Intermittent and Pulsating Impinging Jet

Jundika C. Kurnia
Universiti Teknologi PETRONAS, Bandar Seri
Iskandar, Perak Darul Ridzuan, Malaysia

Peng Xu
China Jiliang University, Hangzhou, Zhejiang, China

Agus P. Sasmito
McGill University, Montreal, Quebec, Canada

Arun S. Mujumdar
McGill University & Western University,
Montreal, Quebec, Canada

CONTENTS

5.1   OVERVIEW OF IMPINGING JET AND IMPINGEMENT DRYING

Over the last decades, impinging jet has received considerable attention with regards to their intricate flow characteristics and their important influence on the transport processes. Despite its apparently simple geometry, the characteristics of the impinging jet flow are complex and have been the subject of numerous studies. Depending on the number of impinging jets (single or multiple) and the design of the nozzle orifices, the structure of an impinging jet can vary considerably. Circular nozzle generates an axisymmetric velocity profile while slot diffuser produces a wide, thin jet with a planar, two-dimensional velocity profile (Nastase and Bode 2018). In general, the flow structure of an impinging jet is constructed by three flow regions, i.e. the free jet region, the stagnation/deceleration region and the wall jet region. These flow regions are illustrated in Figure 5.1.

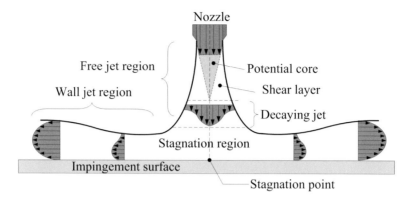

FIGURE 5.1   Schematics of a typical single impinging jet flow.

As can be seen, the free jet region is the region that exists between nozzle exit and the region before impingement (approximately 1.5 diameters from the impingement surface). In this region the flow is mostly unperturbed by the presence of the impingement wall, thus it behaves as a free jet. Here, the velocity gradient in the jet induces a shearing at the edges which transfers momentum laterally outward and entrains the surrounding fluids into the jet, increasing the mass flow of the jet. As a result, the velocity of the jet decreases along the centerline of the jet while the velocity profile is expanding laterally. This creates a non-uniform radial velocity profile within the jet where a quasi-conical, unperturbed region exists in the center of the jet (referred to as the potential core) and a shear layer presents around the potential core. The flow in the potential core is generally low turbulent flow with mean velocity and total pressure similar to the jet exit velocity, while the flow in the shear layer is generally high turbulent with low mean velocity due to its interaction with and entrainment of the surrounding fluid (Carlomagno and Ianiro 2014). Ahead of the free jet region is the stagnation/deceleration region, where the jet flow hits the stagnation point and is deflected radially outward. Away from the stagnation point, the velocity of the jet flow will decrease as it enters a wall jet region where the flow moves laterally outward parallel to the impingement wall and its profile becomes similar to a wall flow.

In most applications, a single jet impingement may not be sufficient; as a result, a series of jets are utilized to achieve the intended objective. The flow characteristic of multiple jet flow is more complicated as compared to the single jet flow due to several factors, i.e., interaction between one jet and the others, collision of the wall jet producing up wash, cross flow and array configuration (Penumadu and Rao 2017). The simplified schematic of the flow structure of multiple impinging jets is presented in Figure 5.2. Here, additional features of the flow can be observed, i.e., the fountain region, which is created due to the combination of two adjacent wall jets, and the secondary stagnation point within the fountain region (Hewakandamby 2009). It should be noted that, depending on the distance between the nozzle and the impingement surface, some of the previously stated regions may not exist.

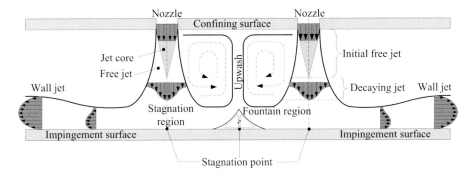

FIGURE 5.2   Schematics of a multiple impinging jet flow.

Impinging jet has been widely used to increase heat transfer between the imping-ing fluid and the impingement surface. Numerous studies have indicated that very high transfer rate can be achieved by utilizing impinging jet, especially in the region around the stagnation point. It is considered one of the highest heat transfer rates for single phase flow (Penumadu and Rao 2017). In addition, the impinging jet con-figuration offers the possibility of pinpointing the location of interest and its desired heat transfer rate (Xu et al. 2012). It is therefore not surprising that impinging jet has been widely adopted in many industrial applications, e.g., electronic chip cooling, drying of thin substrate (fabric and paper), furnace heating, turbine blade cooling, tempering of glass and metal sheets, food processing, solar air heater, internal combustion cooling, and many other applications that require a high transfer rate in limited space.

For drying applications, impinging jet is commonly used to dry thin and flat products such as tiles, tissue, papers, textile and wood veneer (Mujumdar 2006). Owing to its high heat and mass transfer, impinging jet drying is usually imple-mented in a short period to avoid over heating of the drying materials. In addition, impinging jet drying generally involves a large consumption of energy as com-pared to traditional parallel flow drying. This drying method is also not suitable for materials that are highly sensitive to overheating. Therefore, an intermittent impinging jet has been proposed and investigated. It offers a high transfer rate while minimizing energy consumption and the risk of overheating the drying product. As interest in the intermittent impinging jet is growing, numerous studies on this topic have been conducted and reported. To expedite further development and encour-age wider adoption of this drying method, it is necessary to extract the essential information on its basic mechanism and current development stage of this drying method. In this chapter, the basic mechanism of jet impingement will be briefly explained, followed by a description of the heat transfer of impinging jet. Several available correlations to estimate the heat transfer coefficient will be presented. The effect of intermittency and pulsation to the heat transfer performance of imping-ing jet will also be discussed. In the last section, the application of impinging jet, especially the intermittent and pulsation jet, on thermal drying will be discussed and highlighted.

5.2 IMPINGING JET HEAT TRANSFER

Impinging jet has been the major topic in heat transfer fields owing to its ability to offer a high transfer rate. This high transfer rate between impinging jet and the impingement surface originates from its thin hydrodynamic and thermal boundary layer, especially in the stagnation region (Chauhan et al. 2018). The transfer rate of impinging jet is a complex function of many parameters including the Reynolds number, the Pradtl number, nozzle-to-plate spacing and distance from the stagnation point (Jambunathan et al. 1992). In addition, the heat transfer rate is also a strong function of nozzle geometry, impingement surface geometry, flow confinement, turbulence, recovery factor and dissipation of jet temperature. Among those parameters, the Reynolds number is perhaps the most commonly studied parameter. Classically, impinging jet can be classified into laminar and turbulent depending on the Reynolds number based on the nozzle diameter, i.e., dissipated laminar jet when Re<300, fully laminar jet when 300<Re<1000, transition/semi-turbulent jet when 1000<Re<3000 and fully turbulent jet when Re>3000 (Hrycak 1981).

The main parameter in evaluating heat transfer performance of an impinging jet is the Nusselt number. It is defined as the ratio of convective to conductive heat transfer (Çengel and Ghajar 2014), i.e.

$$Nu = \frac{hd}{k},$$ (5.1)

where d is the flow characteristic length (the diameter of the round nozzle or the width of the slot nozzle) and k is the fluid thermal conductivity. The heat transfer coefficient, h, is a function of wall temperature and heat flux and is given by

$$h = \frac{q_w}{\left(T_w - T_{ref}\right)},$$ (5.2)

where q_w is the wall heat flux, T_w is the wall temperature, T_{ref} is the reference temperature, which can be either the jet temperature, T_j, or the adiabatic wall temperature, T_{aw}, depending on the thermal boundary conditions involved (Han and Goldstein 2006). The adiabatic wall temperature, T_{aw}, is the steady-state temperature of the wall when there is no flow. The wall heat flux can be calculated as

$$q_w = k \left. \frac{\partial T}{\partial y} \right|_{y=0}.$$ (5.3)

As stated previously, the heat transfer of impinging jet is a complex function of various parameters. For dimensional analysis, the heat transfer for impinging jet can be described by the relationship between various involved parameters (Hrycak 1981), i.e.

$$Nu = f(\text{Re, Pr, } Ma, \, r \, / \, d, \, z \, / \, d, \, d_p \, / \, d, \, d_c \, / \, d, \, s_n \, / \, d, \, l_1 \, / \, d, \dots, l_n \, / \, d),$$ (5.4)

where Re is the Reynolds number based on jet velocity and nozzle opening (diameter for circular nozzle or width of slot nozzle), Pr is the Prandtl number of the fluid,

Ma is the Mach number, r is the distance from the stagnation point, z is the nozzle-to-impinging-wall distance, d_p is the diameter of the target plate, d_c is the calorimeter (or heat flux) diameter, s_n is the nozzle-to-nozzle spacing and l_1 to l_n are the additional length scales that may be considered in the study, such as roughness of the impingement wall. Some of these parameters are illustrated in Figure 5.3. Accordingly, the general form of the equation for the Nusselt number for impinging jet is

$$Nu = C\,\mathrm{Re}^a\,\mathrm{Pr}^b\,Ma^c \left(\frac{r}{d}\right)^d \left(\frac{z}{d}\right)^e \left(\frac{d_p}{d}\right)^f \left(\frac{d_c}{d}\right)^g \left(\frac{s_n}{d}\right)^h \left(\frac{l_1}{d}\right)^i \cdots \left(\frac{l_n}{d}\right)^u, \quad (5.5)$$

where C is a constant. In their review, Jambunathan et al. (1992) suggested that the simplest correlation for local heat transfer coefficient for impinging jet should be in the following function,

$$Nu = f(\mathrm{Re},\ r/d,\ z/d,\ \mathrm{Pr}). \quad (5.6)$$

Accordingly, the equation for the Nusselt number becomes

$$Nu = C\,\mathrm{Re}^a\,\mathrm{Pr}^b\,f(r/d,\ z/d). \quad (5.7)$$

It should be noted that, in this correlation, the effect of nozzle geometry, confinement and the turbulence upstream of the jet nozzle is not taken into account. Hence,

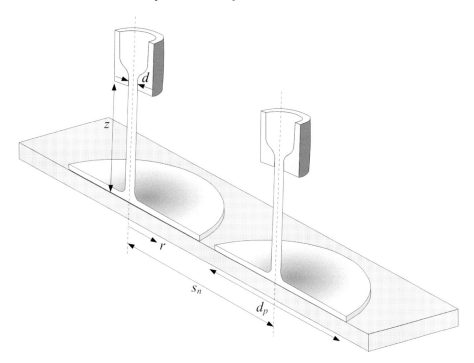

FIGURE 5.3 Important parameters in impinging jet heat transfer.

careful consideration is necessary when using it. Over the last decades, numerous studies on impinging jet have been conducted and a vast amount of heat transfer correlations have been proposed. Some of the available correlations for the Nusselt number of impinging jet are summarized in Table 5.1.

Although impinging jet has high transfer rates, significant efforts are still being devoted to increasing its transfer rate even further. One of the proposed methods to increase transfer rate of impinging jet is by introducing flow pulsation or intermittency. Introducing pulsation to impinging jet is found to disturb the boundary layer and as a result change the local transfer rate. A large vertice is observed close to the jet exit when pulsation is introduced, enlarging the potential core increasing entrainment and in turn widening the area of high heat transfer around the stagnation zone (Farrington and Claunch 1994).

Depending on the flow geometry and conditions, implementation of pulsation or intermittency on the impinging jet can either significantly enhance or deteriorate the resulting heat transfer rate. Significant enhancement has been reported by Zumbrunnen and Balasubramanian (1995), Mladin and Zumbrunnen (1997) and Sailor et al. (1999). These studies reported increases of the Nusselt number by more than 50% due to changes in flow structures, turbulence levels and nonlinear dynamic effects. In some studies, no enhancement or even deterioration of impinging jet heat transfer has been recorded and reported due to the introduction of intermittency. Azevedo et al. (1994) reported up to 20% reduction in heat transfer of pulsed impinging jet as compared to the steady jet counterpart. Similar findings were reported by Sheriff and Zumbrunnen (1994) for a sinusoidal pulsed profile, where up to 17% reduction on the average Nusselt number when the pulse magnitude was large. Pakhomov and Terekhov (2015) reported a reduction in heat transfer enhancement with an increase of the Reynolds number in the range of low frequencies of the pulse-impinging jet. Hofmann et al. (2007) experimentally investigated the effect of pulsation on the flow structure and heat transfer. It was found that the jet becomes broader and the core jet length becomes smaller with pulsation. They reported the significant contribution of frequency and nozzle-to-plate distance in determining the heat transfer enhancement. Moderate frequency and large nozzle-to-plate distance results in reduction of heat transfer in the stagnation region, while high frequency and small nozzle-to-plate distance yields heat transfer enhancement. Esmailpour et al. (2015) investigated the heat transfer performance of a single confined pulsating turbulent impinging. They found that pulsation does increase heat transfer in the wall jet region but at the same time reduces the heat transfer in the stagnation zone. In general, it is concluded that the heat transfer performance of an oscillating impinging jet is enhanced by an increase in the frequency and amplitude of oscillation as well as a decrease in nozzle-to-plate distance. Recently, Zargaradabi et al. (2018) reported that introduction of a sinusoidal pulsation on an impinging jet directed to an asymmetrical concave surface offer positive effect on heat transfer performance. An increase in the time-averaged Nusselt number of up to 8% higher than that of a steady jet was recorded when pulsation with frequency of 40–160 Hz was introduced to the impinging jet.

Another parameter affecting the heat transfer of intermittent impinging jet is the waveform. The most commonly investigated waveform implemented for impinging

TABLE 5.1
Nusselt Number Correlations of Impinging Jet

Nozzle	Nusselt Number Correlation	Remarks	References
Single circular	$Nu = K \, Re^{0.7} \, Pr^{0.33}$	$z/d > 8$, value of K depends on the angle of impingement	Perry (1954)
Single circular	$Nu = C \, Re^{1/3} \exp(-0.037 z/d)$	$C \sim D^{3/2}$, $z/d > 10$, $Re \leq 60,000$	Thurlow (1954)
Single circular	$Nu_0 = 0.023 \, Re^{1/3} \, Re^{0.87}$	$1 < z/d < 10$, $Re \leq 10,000$	Huang (1963)
	$Nu = 0.0181 \, Re^{1/3} \, Re^{0.87}$	Re calculated based on impact velocity	
Single circular	$Nu_0 = 13 \, Re^{1/2} (d/z)$	$z/d > 20$, $Re > 14,000$	Gardon and Cobonpue (1962)
Single circular	$Nu_0 = 1.160 \, Re^{0.447} \, Pr^{0.333}$ for $z/d \leq 7$	$Re \leq 67,000$ Re_a calculated based on arrival velocity	Chamberlain (1966)
	$Nu_0 = 0.384 \, Re_a^{0.569} \, Pr^{0.333}$ for $z/d > 7$		
Single circular	$Nu = \dfrac{Pr^{0.42} \, d \left(1 - 1.1 (d/r)\right)}{r \left(1 + 0.1 \left((z/d) - 6\right) d/r\right)} F$	$2,000 < Re < 400,000$ $2.5 \leq r/d \leq 7.5$ $2 \leq z/d \leq 12$ $0.004 \leq f \leq 0.04$ f is relative nozzle area	Martin (1977)
	$F = \begin{cases} 1.36 \, Re^{0.574} & 2,000 < Re < 30,0000 \\ 0.54 \, Re^{0.667} & 30,000 < Re < 120,0000 \\ 0.151 \, Re^{0.775} & 12,000 < Re < 400,0000 \end{cases}$		
Single circular	$Nu = \dfrac{Re^{0.6}}{\left[A + B(r/d)^n\right]}$	$34,000 < Re < 121,300$	Goldstein and Behbahani (1982)

z/d	A	B	n
6	3.329	0.273	1.3
12	4.577	0.4357	1.14

(Continued)

TABLE 5.1 (Continued)

Nusselt Number Correlations of Impinging Jet

Nozzle	Nusselt Number Correlation	Remarks	References
Single circular	$Nu = \dfrac{24 - \lvert (z/d) - 7.75 \rvert}{533 + 44(r/d)^n} \mathrm{Re}^{0.76}$ Constant wall heat flux: $n = 1.285$ Constant wall temperature: $n = 1.394$	$61{,}000 < \mathrm{Re} < 124{,}000 \ 6 < z/d < 12$	Goldstein et al. (1986)
Single circular	$Nu_0 = 0.15\,\mathrm{Re}^{0.701}\,(z/d)^{-0.25}$ for $10 \le z/d \le 16.7$ and $4{,}860 \le \mathrm{Re} \le 15{,}300$ $Nu_0 = 0.17\,\mathrm{Re}^{0.701}\,(z/d)^{-0.182}$ for $20 \le z/d \le 25$ and $4{,}860 \le \mathrm{Re} \le 15{,}300$ $Nu_0 = 0.388\,\mathrm{Re}^{0.696}\,(z/d)^{-0.345}$ for $6 \le z/d \le 58$ and $6{,}900 \le \mathrm{Re} \le 24{,}900$ $Nu_0 = 0615\,\mathrm{Re}^{067}\,(z/d)^{-0.38}$ for $9 \le z/d \le 41.4$ and $7{,}240 \le \mathrm{Re} \le 34{,}500$	$4{,}860 \le \mathrm{Re} \le 34{,}500 \ 6 \le z/d \le 41$	Mohanty and Tawfek (1993)
Single circular	$Nu = 0.453\,\mathrm{Re}^{0.691}\,\mathrm{Pr}^{1/3}\left(\dfrac{z}{d}\right)^{-0.22}\left(\dfrac{r}{d}\right)^{-0.38}$	$3{,}400 \le \mathrm{Re} \le 41{,}000$ $6 \le z/d \le 58$ $2 \le r/d \le 30$	Tawfek (1996)
Single circular	$Nu = 0.442\,\mathrm{Re}^{0.696}\,\mathrm{Pr}^{1/3}\left(\dfrac{z}{d}\right)^{-0.20}\left(\dfrac{r}{d}\right)^{-0.41}$	$750 \le \mathrm{Re} \le 27{,}000$ $3 \le z/d \le 16$ $0 \le r/d \le 7.14$	Wen and Jang (2003)

Single slot	$Nu_0 = 1.2\,Re^{0.58}\,(z/w)^{-0.62}$ for $2000 < Re \le 50,000$ and $14 < z/w \le 60$ $Nu = 0.36\,Re_a^{0.629}$ for $Re_a > 10,000, Re > 2000, r/w > 6$, and $z/w > 8$	Re_a calculated based on arrival velocity, w is the slot width	Gardon and Akfirat (1966)
Single slot	$Nu = \dfrac{1.53\,Pr^{0.42}\,Re^a}{(x/2w)+(h/2w)+1.39}$ $a = 0.695 - \dfrac{1}{(x/2w)+(h/2w)^{1.33}+3.06}$	$3,000 < Re < 90,000$ $2 \le x/2w \le 25$ $2 \le z/2w \le 10$ $0.01 \le f \le 0.125$ f is relative nozzle area	Martin (1977)
Single slot	$Nu = Re^{0.76}\,Pr^{0.42}\left[a+b\left(\dfrac{z}{d}\right)+c\left(\dfrac{z}{d}\right)^2\right]$ $a = \left[506+13.3(r/d)-19.6(r/d)^2+2.41(r/d)^3 -0.0904(r/d)^4\right]\times10^{-4}$ $b = \left[32-24.3(r/d)-6.536(r/d)^2-0.694(r/d)^3 +0.02574(r/d)^4\right]\times10^{-4}$ $c = (-3.85\times10^{-4})\left(1.147+(r/d)\right)^{-0.0904}$	$6,000 < Re < 60,000$ $1 \le z/w \le 12$ w is the slot width	Huang and El-Genk (1994)

(Continued)

TABLE 5.1 (Continued)

Nusselt Number Correlations of Impinging Jet

Nozzle	Nusselt Number Correlation	Remarks	References
Single slot	$Nu = C\left(\dfrac{z}{w}\right)^a \left(\dfrac{d_t}{w}\right)^b Re^m Pr^n$ Constant: <table><tr><td>z/w</td><td>2–8</td><td>8–12</td></tr><tr><td>C</td><td>0.0516</td><td>0.0803</td></tr><tr><td>a</td><td>0.179</td><td>−0.205</td></tr><tr><td>b</td><td>0.214</td><td>0.162</td></tr><tr><td>m</td><td>0.753</td><td>0.8</td></tr><tr><td>n</td><td>0.4</td><td>0.4</td></tr></table>	Cylindrical target with diameter d_t, $1 \leq d_t / w \leq 4$ $4{,}000 < Re < 20{,}000$	Gori and Bossi (2003)
Single slot	$Nu_0 = 0.815\,Re^{0.55}\,Pr^{0.4}\,(l/w)^{-0.004}$ $Nu = 0074\,Re^{0.775}\,Pr^{0.4}\,(l/w)^{0.182}$	$100 \leq Re \leq 500$	Sexton et al. (2018)
Single	$Nu_0 = 0.72\,Re^{0.5}\,Pr^{0.4}$ for $r/d \leq 0.5$ $Nu = 0.12\,Re^{0.67}\,Pr^{0.4}$ for $r/d = 3$	Jet geometry is not specified. Power for Pr ranging from 0.67 to 0.7.	Specht (2014)
Array circular	$Nu = 0.283\,Re_a^{0.625}$	$z/d > 1$ Re_a calculated based on arrival velocity	Gardon and Cobonpue (1962)
Array circular	$Nu = 0.324\,Re^{0.658}\,(s_n/d)^{-0.342}$	$z/d < 8$	Hrycak (1981)

Row circular	$$Nu = \dfrac{2.9\,Re^{0.7}\exp\left(-0.09(z/d)^{-1.4}\right)}{22.8 + (s_n/d)(z/d)^{1/2}}$$	$10,000 \leq Re \leq 40,000$ $s_n/d = 4,8$ $2 \leq z/d \leq 8$ $0 \leq r/d \leq 6$	Goldstein and Seol (1991)
Array circular	$$Nu = 0.3285\,Re^{0.71}\,Pr^{1/3}\left(\dfrac{z}{d}\right)^{-0.123}\left(\dfrac{s_n}{d}\right)^{-0.725}$$	$3,400 \leq Re \leq 20,500$ $4 \leq s_n/d \leq 8$ $0.25 \leq z/d \leq 6$	Huber and Viskanta (1994)
Array slot	$$Nu = 0.36\,Re_a^{0.629}$$	$6,000 \leq Re_a \leq 600,000$ $Re > 2000$ $16 \leq s_n/w \leq 64$ $z/w > 8$	Gardon and Akfirat (1966)
Array slot	$$Nu = Pr^{0.42}\,\frac{2}{3}\,f_0^{0.75}\left(\dfrac{2Re}{f/f_0 + f_0/f}\right)^{2/3}$$ $$f_0 = \left(60 + 4\left(\dfrac{z}{2w} - 2\right)^2\right)^{-1/2}$$	$1,500 \leq Re \leq 40,000$ $1 \leq z/2w \leq 40$ $0.008 \leq f \leq 2.5 f_0$	Martin (1977)

f is the ratio of the nozzle exit cross section to the area of the square or the hexagon attached ($\pi d_z/4 A_{square}$)

Note: Nu represents the average Nusselt number while Nu_0 represents the Nusselt number at stagnation point.

jet is sinusoidal and square wave pulsation. Xu et al. (2012) numerically investigated the effect of intermittency on heat and mass transfer of multiple impinging jets. The results indicated that only symmetrical pulsation with phase-angle difference offers significant enhancement on the convective heat transfer rate around the secondary stagnation point while pulsation with no phase-angle difference offers marginal heat transfer enhancement. In addition, they found that intermittent (rectangular) pulsating impinging jets provide better heat transfer enhancement compared with that of the sinusoidal case. This result is consistent with the finding by Sheriff and Zumbrunnen (1994) who reported enhancement on the time-averaged Nusselt number by up to 33% when square wave pulsation is introduced. Mohammadpour et al. (2014) conducted a study to determine the optimum arrangement of steady and pulsating impinging multiple submerged slot jets. The investigated jets were steady jet, intermittent (on/off) or sinusoidal jets. The results suggest that the most uniform and enhanced Nusselt number distribution for the array of four jets can be achieved by having a combination of pulsed jets with steady jets. Meanwhile, combinations of intermittent-steady jets offer significantly higher convective heat transfer as compared to sinusoidal-steady arrangement. Ghadi et al. (2016) examined the heat transfer enhancement potential of a pulsed impinging jet. They implemented sinusoidal and step signals to the inlet of the impinging jet and found that a time-varying impinging jet yields an increase in jet development, and its cross section with the wall, and in turn a more uniform Nusselt profile, as compared to the steady jet. In addition, it was observed that a step signal has more significant effect on heat transfer than a sinusoidal pulsation. Recently, Zhang et al. (2018) investigated the effect of pulsation waveforms on the heat transfer of impinging jet. Three different pulsation waveforms were investigated: sinusoidal, rectangular and triangular. The results revealed that the heat transfer rate of the triangular wave is better than that of the sinusoidal wave, which is better than that of the rectangular wave when the Strouhal (St) number is close to the critical point. However, the heat transfer rate of a rectangular wave is better than that of sinusoidal wave, which is better than that of triangular wave when St < 0.06. Farahani and Kowsary (2018) studied the heat transfer of pulsating impinging slot jet. Both the sinusoidal and square wave profile for inlet velocity were investigated. They reported higher heat transfer enhancement for sinusoidal pulsation as compared to square wave pulsation.

In addition to pulsation frequency and waveform, duty cycle has been one of the key factors investigated in the pulsation impinging jet heat transfer. Sailor et al. (1999) studied the effect of duty cycle (the ratio of pulse cycle on-time to total cycle) on the heat transfer performance of an impinging air jet. Their results revealed that duty cycle is crucial in dictating the heat transfer enhancement of a pulsed impinging jet. A similar study was conducted by Marcum et al. (2015) by experimentally investigating the effect of jet location and duty cycle on the fluid mechanics of an unconfined free jet and its heat transfer on an impinging plate. They reported that the duty cycle over a pulse period has a larger impact on the heat transfer of an impinging air jet than that of pulse frequency. This finding is consistent with the result of similar study by Ekkad et al. (2006) which found that the effective blowing ratio due to lowering of the duty cycle at a given blowing ratio seems to play a more important role in combination with pulsing than the pulsing frequency.

As can be inferred from those studies, several factors play important role in determining the heat transfer performance when dealing with time-varying impinging jet. These factors include frequency, amplitude, Reynolds number, pulsation wave form and duty cycle. The effect of pulsation and intermittence on impinging jet heat transfer has been reported to vary ranging from significantly enhance it to marginal or no enhancement. Some even reported adverse effect of pulsation to the heat transfer performance on certain condition. To ensure successful implementation of this method to a specific application, further studies are necessary by taking into consideration the conditions that may be faced during the practical operation. In addition, more studies will provide comprehensive figures on how these various key parameters affecting the flow and heat transfer of impinging jet and provide guideline in designing it for real applications.

5.3 IMPINGING JET DRYING

During drying, simultaneous heat and mass transfer occurs within the drying materials and their surroundings. Thermal energy will be transferred from the surrounding environment into the drying materials through convection, radiation and conduction while moisture is driven from the drying material to the surrounding via diffusion, evaporation and convection. Consequently, to shorten drying time or improve drying rate, either heat transfer rate or mass transfer rate or both transfer rates should be improved. As discussed in the previous section, impinging jet offers enhancement to both mass and heat transfer. As such, impinging jet has been commonly adopted in convective thermal drying, especially for continuous and thin material, which requires short drying time. Figure 5.4 depicts the basic principles of the drying process using impinging jet.

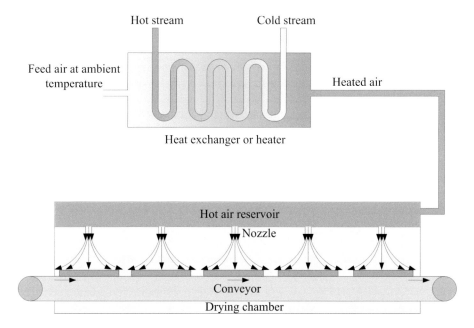

FIGURE 5.4 Basic principle of impinging jet drying.

In comparison with conventional parallel drying, impinging jet drying offers several advantages (Xiao and Mujumdar 2014), i.e., shorter drying time due to higher heat and mass transfer, less energy consumption due to shorter drying time and better quality product since the product is not exposed to a prolonged drying condition which can degrade and discolor the final product. Several studies on the application of impinging jet in drying process of a thin porous material such as fruit slice, vegetable slice, herbs, meats, seeds, textiles, papers, woods, ceramics and concrete slabs have been conducted and reported.

Wählby et al. (2000) experimentally studied the impact of directed forced convection (impingement) on food quality. Yeast buns (representing category bread) and pork cutlet (representing meats) were selected as samples in their study. They reported a shorter cooking time and more uniform browning for meat. As for the bun, cooking time was essentially similar to a traditional oven, but at a much lower air temperature setting. Banooni et al. (2008) evaluated the heat transfer and product quality aspects of an impinging jet drying for flat breads. Better baking conditions, e.g., good browning, uniform color and high volume increase were achieved with the impingement oven in comparison with the conventional direct fired counterpart. In addition, a higher jet velocity at lower oven temperatures yield shorter baking times for the same quality product. They highlighted that to achieve all these desirable results, a well-designed and operated impingement oven is compulsory. Similar findings were found for drying Spirulina platensis, where impinging jet drying with high jet velocity but low temperature is desirable to ensure high quality product and lower energy consumption (Yamsaengsung and Bualuang 2010).

Bórquez et al. (1999; 2008) investigated and developed an impinging jet dryer for pine sawdust and mackerel press-cake. Shorter residence times and higher gas temperatures was found to increase drying rates and heat and mass transfer coefficients. High moisture removal and high quality dried mackerel press-cake product (with minimum losses of the valuable omega-3 fatty acids) was reported for impinging drying with superheated steam. Tapaneyasin et al. (2005) implemented a jet-spouted bed dryer for shrimp and evaluated the effects of various parameters including size of shrimp, level and pattern of inlet drying air temperature on drying characteristics and quality of the product. Constant inlet air temperature of 100°C was found to produce the best product quality in terms of low percentage of shrinkage, high percentage of rehydration, low maximum shear force and high value of redness compared to shrimp dried using other conditions.

Xiao et al. (2010a) evaluated the effects of air temperature, air velocity and air relative humidity on air-impingement drying characteristics of carrot cubes and the quality of dried carrots. Drying air temperature was found to be the dictating parameter on drying time of carrot cubes followed by air relative humidity and air velocity. In addition, they found that the rehydration ratio of dried carrot cubes was significantly affected by drying air temperature and air relative humidity. In their other study, Xiao et al. (2010b) dried Monukka seedless grapes by using impinging jet drier. They evaluated the effect of drying temperature and air velocity on the drying characteristics and the quality of dried grapes in terms of texture and vitamin C content. Drying temperature has a bigger role in determining the drying time than the air velocity. Meanwhile, the drying temperature was the major factor controlling the retention of vitamin C, while there was no direct correlation between air velocity and vitamin C retention. In addition, the hardness of dried grapes increased as the drying temperature increased. Another report by Xiao et al. (2014, 2015) highlighted the applicability of hot air impingement drying to dry American ginseng slices. In their studies, they investigated the effect of drying temperature, air velocity and sample thickness on thin-layer air-impingement drying characteristics and quality of American ginseng slices in terms of color parameters, ginsenosides content, rehydration ratio and microstructure. Drying time was found to be principally affected by drying temperature followed by sample thickness and air velocity. Based on quality evaluation, the change of color is mainly affected by drying temperature and sample thickness while air velocity exhibits no significant effect. In their review, Xiao and Mujumdar (2014) highlighted that impinging drying offers significantly shorter drying time as compared to thin-layer solar drying and traditional convective drying for the same material.

Impinging drying has also been adopted by Zhang et al. (2011) to dry Hami melon flakes. They evaluated the effect of drying temperature and air velocity on the drying characteristics. A drying time of 5–9 h can be achieved, which is significantly shorter than natural sun drying that requires around 10 days to complete. In their following study (Zheng et al. 2014), they attempted to reduce the drying time even further by incorporating the middle short-wave infrared radiation to their impinging dryer. A drying time of about 2–3.5 h can be achieved, which is the fastest among the considered drying technologies. Drying can be operated easily using the middle short-wave radiation combined with air-impingement drying technology since the energy needed to trigger drying is very small.

The effect of air-impingement drying on the drying characteristics and quality of tortilla chips was examined by Lujan-Acosta et al. (1997a; 1997b). They reported several important findings: i) the drying rate was mostly affected by the air temperature; ii) the texture was crispier at higher air temperatures; iii) the shrinkage of the piece was higher at a lower convective heat transfer coefficient; and iv) the microstructure looked smoother at a higher air temperature. A combination of air-impingement drying and frying is suggested to produce low-fat tortilla chips with good flavour and texture. Instead or air-impingement drying, some researchers used superheated steam to replace the hot air in impingement drying of tortilla chips. Li et al. (1999) investigated superheated steam impingement drying effects on product quality such as shrinkage, crispiness and microstructure and evaluated the effect of steam temperature and flow rate on the drying characteristics. The effect of steam temperature was found to be more pronounced on drying characteristics than the heat transfer coefficient. It also has a more significant effect on product qualities such as texture, microstructure and shrinkage of tortilla chips during drying. At elevated temperatures, the superheated steam provided higher drying rates than air-impingement drying. However, steam impingement drying produces more starch gelatinization that its air-impingement counterpart. A similar study was conducted by Caxieta et al. (2002), who implemented superheated steam impingement drying for potato chips. They evaluated the effect of superheated steam temperature and convective heat transfer coefficient on the drying rate and product quality attributes (shrinkage, density, porosity, color, texture and nutrition loss) of potato chips. It was found that the temperature and convective heat transfer coefficient had a significant effect on the drying rate during superheated steam impingement drying.

Jung et al. (2015) evaluated the feasibility of impingement drying as a rapid drying method to dry wet apple pomace (WAP) and concluded that impingement drying is a fast and effective method for preparing dried apple pomace flour (APF) with highly retained bioactive compounds, and apple pomace-fortified products maintained or even had improved quality. Li et al. (2015) investigated the effects of air-impingement jet drying on the drying kinetics, nutrient retention and rehydration characteristics of onion (Allium cepa) slices. The results revealed that air-impingement jet offers shorter drying time and better quality of products, indicating impinging jet drying can be widely applied to food drying process. In their later study, Li et al. (2016) conducted an evaluation of the drying kinetics, heating characteristics and energy consumption of heat pipe-impingement drying (HID) of kiwi fruit slices. They found that low temperature and high air velocity provides a better drying option. It produces high quality dried kiwifruit and consumes less energy. Evaluation of potential energy savings by using impinging jet drying was conducted by Li et al. (2013). They estimated that a potential saving of up to 12% can be achieved under the investigated situation. It should be noted, however, that their study is focusing on the application of impinging jet in the bread baking process.

For seed product, impinging stream drying is used instead of conventional impingement drying. Pruengam et al. (2016) conducted an experimental study to investigate the evolution of mechanical properties of parboiled brown rice kernels during impinging stream drying. Effects of the drying temperature and tempering periods were evaluated. The result indicated that the effect of the drying temperature and existence of tempering on the mechanical properties of the kernels are strongly dependent on the kernel moisture contents immediately after drying. Wachiraphansakul and Devahastin (2005) implemented a jet-spouted dryer to dry okara (a soy residue obtained from soymilk production). Effects of several parameters, i.e. inlet air velocity, inlet air temperature, initial bed height and heating duration, on the drying kinetics and quality of the final product were evaluated. The results indicated that all drying conditions considered in this study were able to reduce the amount of urease to an acceptable level. In addition, increasing the heating duration, air velocity, and hot air temperature resulted in a significantly higher rate of reduction of urease activity.

Other than food product, impingement drying has also been investigated for other materials. Ratanawilai et al. (2015) compared impinging hot air drying and microwave heating to dry rubberwood slabs. The investigation revealed the superiority of impinging jet, which offers significantly shorter drying time than microwave heating. It is worth noting that both methods are practical without generating post-drying residual stresses inside the rubberwood. Chen et al. (1995) presented a complete drying history of paper dried under an array of multiple round jets. They proposed general correlations of three parameters, i.e., the constant drying rate, the critical moisture content and the exponent for the power law falling rate period relationship to predict drying time for a wide range of drying conditions. Bond et al. (1994) utilized impinging jet superheated steam for paper drying. It was found that over a wide range of operating conditions, the drying of paper by impinging jets of superheated steam proceeds by a constant rate period followed by a falling rate period. Francis and Wepfer (1996) investigated jet impingement drying of semi-porous textile composites. The increased heat flux at the surface was found to significantly affect the internal transport phenomena in the porous material. Another finding was that the increased temperature and gas phase pressure gradients in the porous solid during the wet region drying period (initial stages of drying) increase the thermal convection terms in the conservation of energy equation during this drying regime.

In tandem with experimental investigation, several numerical investigations on impinging jet drying have been conducted and reported. De Bonnis and Ruocco (2014) developed an axisymmetric transfer phenomena model to investigate the transient behaviour of temperature and moisture content within a moist protrusion under an air impinging jet. It was highlighted that impinging jet can be employed to induce a desired superficial finish. Khomwachirakul et al. (2016) utilized a combination of computational fluid dynamics and discrete element method to predict, for the first time, the multiphase transport phenomena within a coaxial of a coaxial impinging steam dryer. They highlighted that the CFD-DEM model could be utilized to predict the particle motion behaviour and led to more physically realistic results than the CFD model. Ljung et al. (2017) conducted numerical study of an impingement jet dryer with a total of 9 pairs of nozzles that dries sheets of metal. They highlighted that the evaporation rate in the impingement dryer is highly dependent on the saturation of vapour in the inlet air.

Despite their impressive drying performance, impinging jet drying has certain inherent drawbacks, i.e., higher energy consumption for the same drying period as compared to the traditional parallel flow drying, it may not be suitable for highly sensitive material due to overheating issues and the higher possibility of burning product. As highlighted at the beginning of this section, drying is delicate process that involves simultaneous heat and mass transfer. The moisture content is transferred from the drying material to the surroundings through diffusion, evaporation and convection. Increasing drying air velocity and temperature will increase drying rate and thus shorten drying time if the water content is available at or near the surface of the drying material. If water content is located deep inside the drying material, continuously applying high velocity and high temperature jet to the material will not expedite the drying process. In fact, it may even lead to quality degradation and damage to the product surface. To avoid this problem, it is essential to match the heat supply with the heat required for drying to achieve optimum drying process, i.e., a high drying rate, minimum energy consumption and high quality product. To achieve this, the heat supplied for the drying process should be delivered periodically or sequentially according to the drying kinetics that govern the overall process through intermittency or pulsation. Intermittent drying can be conducted by varying air flow rate, air temperature, air relative humidity, operating pressure or heat transfer mode, e.g., by sequentially applying conduction, convection or radiation (Chua et al. 2003). The idea of implementing intermittency on drying is not new. Chua et al. (2001) dried banana slice in a two-stage heat pump dryer capable of producing stepwise control of the inlet drying air temperature while keeping absolute humidity constant. It was found that better quality product in terms of color of the dried product can be obtained by using stepwise air temperature variation. A similar study was conducted by Silva et al. (2015) for the whole banana. Three intermittent processes were performed with the same intermittency ratio (1/2), with intermittency times of 0.5, 1.0 and 2.0 h. Comparison between the curves for continuous and intermittent drying with 1 h of intermittence indicates that savings in time (and, consequently, in energy) can be achieved. Moreover, it is found that intermittency produce less stress within the product than continuous drying.

Silva et al. (2014) experimentally compared continuous and discontinuous drying of sliced pear. Their study was focused on the potential energy saving by discontinuous drying. An analysis of pause periods was performed in order to predict energy savings and to assess the feasibility of using solar energy for this kind of convective drying procedure. Their findings indicate that an increase in the number of pauses can lead to significant energy savings. In their following study (Silva et al. 2016), the whole pear was dried using three-stage intermittent drying. In this study, shrinkage and the variable diffusion coefficient were considered. The intermittency of the process contributed to decrease the total time of convective drying. It should be noted, however, that shorter pause periods are disadvantageous and lead to longer drying time, because the homogenization of the radial moisture distribution is incomplete before starting the next convective drying phase. A review of intermittent drying of porous materials has been reported by Rufino Franco and Barbosa de Lima (2016). They highlight several features of an intermittent drying such as thermal energy savings, shorter effective drying time, higher moisture removal rates, higher moisture at the surface of the product at the end of tempering period and higher product quality represented by lower risk of cracking and brittleness as well as better colour and nutrient retention.

Several studies on pulsation and intermittent impinging jet drying have been conducted and reported. Patterson et al. (2003) investigated the application of pulse combustion-based jet impingement to paper. Pulse combustion impingement was found to produce significantly higher drying rates in comparison to steady flow impingement. It should be noted, however, that the performance depends on the pulse combustor configuration, the impingement nozzle geometry, the substrate supporting the sheet, and the presence or absence of vacuum. Hagadorn (2005) investigated the increase in heat transfer and drying rates by using pulse combustion impingement drying. It was concluded that combustion impingement drying offers an improvement over the steady impingement drying currently in commercial use. It has higher heat transfer rates and a lower impact on the environment. Liewkongsataporn et al. (2006) studied the effects of impingement condition on impingement drying of paper samples by using a small pulse combustor. It was found that some drying enhancement can be achieved by the pulse impingement as compared to a steady jet impingement. The pulsating jet was restricted to one condition due to the limitation of pulse combustor construction. Similarly, Wu et al. (2006) experimentally studied pulsating jet impingement drying on paper samples with various flow rates through a pulse combustor. Unfortunately, no comparison with steady jets was reported. Lihong et al. (2011) designed and evaluated a pulsed air-impingement dryer with large processing capacity to replace a conventional air-impingement dryer. Their study indicated that, as compared to conventional air-impingement, the dehydration intensity and processing capacity of the pulsed air-impingement dryer was higher. Ahrens et al. (2011)

numerically investigated the design tradeoffs between heat transfer rate (i.e., drying rate), heat transfer (jet thermal energy) effectiveness and heat transfer enhancement due to the pulsating jet. It was observed that at equal jet thermal energy effectiveness around 2.4 times greater average heat flux (drying rate) can be achieved with pulsating impingement than with steady impingement.

Kurnia et al. (2013) reported the effect of jet velocity, pulsation and intermittency on the jet inlet; slab geometry and slab thickness were investigated. An important highlight from this study is that lower energy consumption can be achieved by impinging jet with pulsating and/or intermittent flow while offering comparable drying kinetics as that of a steady jet, which shows potential for energy saving. On their following study (Kurnia et al. 2017), the effect of inlet temperature intermittency and inlet velocity intermittency on drying characteristics and energy consumption were evaluated. The inlet temperature intermittency is conducted by alternately raising inlet temperature to drying temperature and lowering it to the ambient temperature at a certain period, while the velocity intermittency is achieved by alternately supplying the hot air to several drying chambers. Four different configurations were simulated and evaluated, i.e., one-, two-, three- and four-chamber configurations were evaluated. The results reveal that the drying rate of intermittent jet is inferior as compared to the steady impinging jet counterpart under the same inlet conditions. In addition, it was found that as the number of drying chambers increases, the drying rate goes down. It is worth emphasizing, however, that the intermittent impinging jet drying offers advantages in terms of temperature uniformity and energy conservation. For the same energy usage, the production rate of single-chamber drying configuration is only one fourth that of the four-chamber configuration. This indicates the potential of multi-chamber configuration in a real drying application.

5.4   CONCLUDING REMARKS

Owing to its high transfer rate, impinging jet has been adopted in various industrial applications that require high transfer rate in limited space. Its intricate flow behaviour and corresponding transport process has attracted significant attention from academia and industry. Accordingly, numerous studies have been conducted, various parameters and impinging jet configurations have been evaluated and a range of correlations to capture its transfer rate have been proposed. The effects of pulsation and intermittency on the flow characteristics and corresponding transfer performance have been evaluated and reported. On drying application, impinging jet has been commonly applied to dry thin and flat products such as tiles, tissue, papers, textile and wood veneer. To avoid overheating and reduce energy consumption, an intermittent impinging jet has been proposed and investigated. Several studies have confirmed potential energy savings and better product quality resulted from intermittent drying. Unlike intermittence in conventional parallel flow drying, only a limited number of studies on the intermittent impinging jet have been conducted and reported. More studies shall be conducted in the future to gain a more comprehensive knowledge on intermittent impinging drying and in turn to expedite the application of this drying technology.

REFERENCES

Ahrens, F., W. Liewkongsataporn, and T. Patterson. 2011. "Optimizing the Benefits of Pulse Combustion Impingement for Paper Drying." *Chemical Engineering & Technology* 34 (7): 1109–15. doi:10.1002/ceat.201100027.

Azevedo, L. F. A., B. W. Webb, and M. Queiroz. 1994. "Pulsed Air Jet Impingement Heat Transfer." *Experimental Thermal and Fluid Science* 8 (3): 206–13. doi:10.1016/0894-1777(94)90049-3.

Banooni, S., S. M. Hosseinalipour, A. S. Mujumdar, E. Taheran, M. Bahiraei, and P. Taherkhani. 2008. "Baking of Flat Bread in an Impingement Oven: An Experimental Study of Heat Transfer and Quality Aspects." *Drying Technology* 26 (7): 902–9. doi:10.1080/07373930802142614.

Bond, J. F., A. S. Mujumdar, A. R. P. van Heiningen, and W. J. M. Douglas. 1994. "Drying Paper by Impinging Jets of Superheated Steam. Part 1: Constant Drying Rate in Superheated Steam." *The Canadian Journal of Chemical Engineering* 72 (3): 446–51. doi:10.1002/cjce.5450720309.

Bórquez, R. M., E. R. Canales, and H. R. Quezada. 2008. "Drying of Fish Press-Cake with Superheated Steam in a Pilot Plant Impingement System." *Drying Technology* 26 (3): 290–98. doi:10.1080/07373930801897986.

Bórquez, R., W. Wolf, W. D. Koller, and W. E. L. Spieß. 1999. "Impinging Jet Drying of Pressed Fish Cake." *Journal of Food Engineering* 40 (1): 113–20. doi:10.1016/S0260-8774(99)00047-3.

Caixeta, Aline T., Rosana Moreira, and M. Elena Castell-Perez. 2002. "Impingement Drying of Potato Chips." *Journal of Food Process Engineering* 25 (1): 63–90. doi:10.1111/j.1745-4530.2002.tb00556.x.

Carlomagno, Giovanni Maria, and Andrea Ianiro. 2014. "Thermo-Fluid-Dynamics of Submerged Jets Impinging at Short Nozzle-to-Plate Distance: A Review." *Experimental Thermal and Fluid Science* 58 (October): 15–35. doi:10.1016/j.expthermflusci.2014.06.010.

Çengel, Yunus A., and Afshin Jahanshahi Ghajar. 2014. *Heat and Mass Transfer: Fundamentals & Applications*. McGraw-Hill Education.

Chamberlain, John Edward. 1966. "Heat Transfer between a Turbulent Round Jet and a Segmented Flat Plate Perpendicular to It." MSc Thesis, Newark, New Jersey: Newark College of Engineering, New Jersey Institute of Technology. http://archives.njit.edu/vol01/etd/1960s/1967/njit-etd1967-002/njit-etd1967-002.pdf.

Chauhan, Ranchan, Tej Singh, N. S. Thakur, Nitin Kumar, Raj Kumar, and Anil Kumar. 2018. "Heat Transfer Augmentation in Solar Thermal Collectors Using Impinging Air Jets: A Comprehensive Review." *Renewable and Sustainable Energy Reviews* 82 (February): 3179–90. doi:10.1016/j.rser.2017.10.025.

Chen, G., V. G. Gomes, and W. J. M. Douglas. 1995. "Impingement Drying of Paper." *Drying Technology* 13 (5–7): 1331–44. doi:10.1080/07373939508917025.

Chua, K. J., A. S. Mujumdar, and S. K. Chou. 2003. "Intermittent Drying of Bioproducts—an Overview." *Bioresource Technology* 90 (3): 285–95. doi:10.1016/S0960-8524(03)00133-0.

Chua, K. J., A. S. Mujumdar, M. N. A. Hawlader, S. K. Chou, and J. C. Ho. 2001. "Convective Drying of Agricultural Products. Effect of Continuous and Stepwise Change in Drying Air Temperature." *Drying Technology* 19 (8): 1949–60. doi:10.1081/DRT-100107282.

De Bonis, Maria Valeria, and Gianpaolo Ruocco. 2014. "Conjugate Heat and Mass Transfer by Jet Impingement over a Moist Protrusion." *International Journal of Heat and Mass Transfer* 70 (March): 192–201. doi:10.1016/j.ijheatmasstransfer.2013.11.014.

Ekkad, Srinath V., Shichuan Ou, and Richard B. Rivir. 2006. "Effect of Jet Pulsation and Duty Cycle on Film Cooling From a Single Jet on a Leading Edge Model." *Journal of Turbomachinery* 128 (3): 564–71. doi:10.1115/1.2185122.

Esmailpour, Kazem, Mostafa Hosseinalipour, Behnam Bozorgmehr, and Arun S. Mujumdar. 2015. "A Numerical Study of Heat Transfer in a Turbulent Pulsating Impinging Jet." *The Canadian Journal of Chemical Engineering* 93 (5): 959–69. doi:10.1002/cjce.22169.

Farahani, Somayeh Davoodbadi, and Farshad Kowsary. 2018. "Heat Transfer from Pulsating Laminar Impingement Slot Jet on a Flat Surface with Inlet Velocity: Sinusoidal and Square Wave." *Heat Transfer Engineering* 39 (10): 901–13. doi:10.1080/01457632.2017. 1338868.

Farrington, Robert B., and Scott D. Claunch. 1994. "Infrared Imaging of Large-Amplitude, Low-Frequency Disturbances on a Planar Jet." *AIAA Journal* 32 (2): 317–23. doi:10.2514/3.11987.

Francis, Nicholas D., and William J. Wepfer. 1996. "Jet Impingement Drying of a Moist Porous Solid." *International Journal of Heat and Mass Transfer* 39 (9): 1911–23. doi:10.1016/0017-9310(95)00269-3.

Gardon, Robert, and J. Cahit Akfirat. 1966. "Heat Transfer Characteristics of Impinging Two-Dimensional Air Jets." *Journal of Heat Transfer* 88 (1): 101–7. doi:10.1115/1.3691449.

Gardon, Robert, and J. Cobonpue. 1962. "Heat Transfer Between a Flat Plate and Jets of Air Impinging on It." In *International Developments in Heat Transfer*, 454–60. New York, N.Y.

Ghadi, Sina, Kazem Esmailpour, Mostafa Hosseinalipour, and Mehrdad Kalantar. 2016. "Dynamical Study of Pulsed Impinging Jet with Time Varying Heat Flux Boundary Condition." *Heat Transfer—Asian Research* 45 (1): 85–100. doi:10.1002/htj.21154.

Goldstein, R. J., and A. I. Behbahani. 1982. "Impingement of a Circular Jet with and without Cross Flow." *International Journal of Heat and Mass Transfer* 25 (9): 1377–82. doi:10.1016/0017-9310(82)90131-4.

Goldstein, R. J., A. I. Behbahani, and K. Kieger Heppelmann. 1986. "Streamwise Distribution of the Recovery Factor and the Local Heat Transfer Coefficient to an Impinging Circular Air Jet." *International Journal of Heat and Mass Transfer* 29 (8): 1227–35. doi:10.1016/0017-9310(86)90155-9.

Goldstein, R. J., and W. S. Seol. 1991. "Heat Transfer to a Row of Impinging Circular Air Jets Including the Effect of Entrainment." *International Journal of Heat and Mass Transfer* 34 (8): 2133–47. doi:10.1016/0017-9310(91)90223-2.

Gori, F., and L. Bossi. 2003. "Optimal Slot Height in the Jet Cooling of a Circular Cylinder." *Applied Thermal Engineering* 23 (7): 859–70. doi:10.1016/S1359-4311(03)00025-5.

Hagadorn, Charles C. 2005. "Measurement and Mapping of Pulse Combustion Impingement Heat Transfer Rates." Thesis, Georgia Institute of Technology. https://smartech.gatech. edu/handle/1853/7467.

Han, B., and R. J. Goldstein. 2006. "Jet-Impingement Heat Transfer in Gas Turbine Systems." *Annals of the New York Academy of Sciences* 934 (1): 147–61. doi:10.1111/j.1749-6632.2001.tb05849.x.

Hewakandamby, Buddhika N. 2009. "A Numerical Study of Heat Transfer Performance of Oscillatory Impinging Jets." *International Journal of Heat and Mass Transfer* 52 (1): 396–406. doi:10.1016/j.ijheatmasstransfer.2008.07.004.

Hofmann, Herbert Martin, Daniela Luminita Movileanu, Matthias Kind, and Holger Martin. 2007. "Influence of a Pulsation on Heat Transfer and Flow Structure in Submerged Impinging Jets." *International Journal of Heat and Mass Transfer* 50 (17): 3638–48. doi:10.1016/j.ijheatmasstransfer.2007.02.001.

Hrycak, Peter. 1981. "Heat Transfer from Impinging Jets. A Literature Review:" Fort Belvoir, VA: Defense Technical Information Center. doi:10.21236/ADA106723.

Huang, G. C. 1963. "Investigations of Heat-Transfer Coefficients for Air Flow Through Round Jets Impinging Normal to a Heat-Transfer Surface." *Journal of Heat Transfer* 85 (3): 237–43. doi:10.1115/1.3686082.

Huang, Lianmin, and Mohamed S. El-Genk. 1994. "Heat Transfer of an Impinging Jet on a Flat Surface." *International Journal of Heat and Mass Transfer* 37 (13): 1915–23. doi:10.1016/0017-9310(94)90331-X.

Huber, Aaron M., and Raymond Viskanta. 1994. "Effect of Jet-Jet Spacing on Convective Heat Transfer to Confined, Impinging Arrays of Axisymmetric Air Jets." *International Journal of Heat and Mass Transfer* 37 (18): 2859–69. doi:10.1016/0017-9310(94)90340-9.

Jambunathan, K., E. Lai, M. A. Moss, and B. L. Button. 1992. "A Review of Heat Transfer Data for Single Circular Jet Impingement." *International Journal of Heat and Fluid Flow* 13 (2): 106–15. doi:10.1016/0142-727X(92)90017-4.

Jung, Jooyeoun, George Cavender, and Yanyun Zhao. 2015. "Impingement Drying for Preparing Dried Apple Pomace Flour and Its Fortification in Bakery and Meat Products." *Journal of Food Science and Technology* 52 (9): 5568–78. doi:10.1007/s13197-014-1680-4.

Khomwachirakul, Patiwat, Sakamon Devahastin, Thanit Swasdisevi, and Somchart Soponronnarit. 2016. "Simulation of Flow and Drying Characteristics of High-Moisture Particles in an Impinging Stream Dryer via CFD-DEM." *Drying Technology* 34 (4): 403–19. doi:10.1080/07373937.2015.1081930.

Kurnia, Jundika C., Agus P. Sasmito, Wei Tong, and Arun S. Mujumdar. 2013. "Energy-Efficient Thermal Drying Using Impinging-Jets with Time-Varying Heat Input – A Computational Study." *Journal of Food Engineering* 114 (2): 269–77. doi:10.1016/j.jfoodeng.2012.08.029.

Kurnia, Jundika C., Agus P. Sasmito, Peng Xu, and Arun S. Mujumdar. 2017. "Performance and Potential Energy Saving of Thermal Dryer with Intermittent Impinging Jet." *Applied Thermal Engineering* 113 (February): 246–58. doi:10.1016/j.applthermaleng.2016.11.036.

Li, Wenfeng, Moyi Wang, Xiao Xulin, Baoshan Zhang, and Xingbin Yang. 2015. "Effects of Air-Impingement Jet Drying on Drying Kinetics, Nutrient Retention and Rehydration Characteristics of Onion (Allium Cepa) Slices." *International Journal of Food Engineering* 11 (3): 435–446. doi:10.1515/ijfe-2014-0269.

Li, Wenfeng, Li Yuan, Xuling Xiao, and Xingbin Yang. 2016. "Dehydration of Kiwifruit (Actinidia Deliciosa) Slices Using Heat Pipe Combined with Impingement Technology." *International Journal of Food Engineering* 12 (3): 265–276. doi:10.1515/ijfe-2015-0165.

Li, X. -D., M. Alamir, E. Witrant, G. Della-Valle, O. Rouaud, L. Boillereaux, and Ch. Josset. 2013. "Further Investigations on Energy Saving by Jet Impingement in Bread Baking Process." *IFAC Proceedings Volumes*, 5th IFAC Symposium on System Structure and Control, 46 (2): 701–6. doi:10.3182/20130204-3-FR-2033.00017.

Li, Y. B., J. Seyed-Yagoobi, R. G. Moreira, and R. Yamsaengsung. 1999. "Superheated Steam Impingement Drying of Tortilla Chips." *Drying Technology* 17 (1–2): 191–213. doi:10.1080/07373939908917525.

Liewkongsataporn, W., T. Patterson, F. Ahrens, and J. Loughran. 2006. "Impingement Drying Enhancement Using a Pulsating Jet." In *Proceedings of the 15th International Drying Symposium*, edited by István Farkas. Budapest, Hungary.

Lihong, Wang, Gao Zhenjiang, Lin Hai, Hong-Wei Xiao, and Zhang Qian. 2011. "Pulsed Air-Impingement Dryer." *Nongye Jixie Xuebao/Transactions of the Chinese Society of Agricultural Machinery* 42 (October): 141–44.

Ljung, Anna-Lena, L. Robin Andersson, Anders G. Andersson, T. Staffan Lundström, and Mats Eriksson. 2017. "Modelling the Evaporation Rate in an Impingement Jet Dryer with Multiple Nozzles." *International Journal of Chemical Engineering*, 5784627(1–9). doi:10.1155/2017/5784627.

Lujan-Acosta, Javier, and Rosana G. Moreira. 1997b. "Reduction of Oil in Tortilla Chips Using Impingement Drying." *Food Science and Technology* 30 (8): 834–40. doi:10.1006/fstl.1997.0282.

Lujan-Acosta, Javier, Rosana G. Moreira, and Jamal Seyed-Yagoobi. 1997a. "Air-Impingement Drying of Tortilla Chips." *Drying Technology* 15 (3–4): 881–97. doi:10.1080/07373939708917266.

Marcum, W. R., S. R. Cadell, and C. Ward. 2015. "The Effect of Jet Location and Duty Cycle on the Fluid Mechanics of an Unconfined Free Jet and Its Heat Transfer on an Impinging Plate." *International Journal of Heat and Mass Transfer* 88 (September): 470–80. doi:10.1016/j.ijheatmasstransfer.2015.04.041.

Martin, Holger. 1977. "Heat and Mass Transfer between Impinging Gas Jets and Solid Surfaces." In *Advances in Heat Transfer*, edited by James P. Hartnett and Thomas F. Irvine, 13:1–60. Elsevier. doi:10.1016/S0065-2717(08)70221-1.

Mladin, E. C., and D. A. Zumbrunnen. 1997. "Local Convective Heat Transfer to Submerged Pulsating Jets." *International Journal of Heat and Mass Transfer* 40 (14): 3305–21. doi:10.1016/S0017-9310(96)00380-8.

Mohammadpour, Javad, Mohammad Mehdi Zolfagharian, Arun S. Mujumdar, Mehran Rajabi Zargarabadi, and Mohammad Abdulahzadeh. 2014. "Heat Transfer under Composite Arrangement of Pulsed and Steady Turbulent Submerged Multiple Jets Impinging on a Flat Surface." *International Journal of Thermal Sciences* 86 (December): 139–47. doi:10.1016/j.ijthermalsci.2014.07.004.

Mohanty, A. K., and A. A. Tawfek. 1993. "Heat Transfer Due to a Round Jet Impinging Normal to a Flat Surface." *International Journal of Heat and Mass Transfer* 36 (6): 1639–47. doi:10.1016/S0017-9310(05)80073-0.

Mujumdar, Arun S. 2006. *Handbook of Industrial Drying*, Third Edition. CRC Press.

Nastase, Ilinca, and Florin Bode. 2018. "Impinging Jets – a Short Review on Strategies for Heat Transfer Enhancement." *E3S Web of Conferences* 32: 01013. doi:10.1051/e3sconf/20183201013.

Pakhomov, M. A., and V. I. Terekhov. 2015. "Numerical Study of Fluid Flow and Heat Transfer Characteristics in an Intermittent Turbulent Impinging Round Jet." *International Journal of Thermal Sciences* 87 (January): 85–93. doi:10.1016/j.ijthermalsci.2014.08.007.

Patterson, Timothy, F. Ahrens, and G. Stipp. 2003. "High Performance Impingement Paper Drying Using Pulse Combustion Technology." In *TAPPI Spring Technical Conference and Exhibit*, 941–62. https://www.researchgate.net/publication/286127224_High_performance_impingement_paper_drying_using_pulse_combustion_technology.

Penumadu, Prithvi Sai, and Arvind Gangoli Rao. 2017. "Numerical Investigations of Heat Transfer and Pressure Drop Characteristics in Multiple Jet Impingement System." *Applied Thermal Engineering* 110 (January): 1511–24. doi:10.1016/j.applthermaleng.2016.09.057.

Perry, K. P. 1954. "Heat Transfer by Convection from a Hot Gas Jet to a Plane Surface." *Proceedings of the Institution of Mechanical Engineers* 168 (1): 775–84. doi:10.1243/PIME_PROC_1954_168_071_02.

Pruengam, Pimpan, Somchart Soponronnarit, Somkiat Prachayawarakorn, and Sakamon Devahastin. 2016. "Evolution of Mechanical Properties of Parboiled Brown Rice Kernels during Impinging Stream Drying." *Drying Technology* 34 (15): 1843–53. doi:10.1080/07373937.2016.1213277.

Ratanawilai, T., C. Nuntadusit, and N. Promtong. 2015. "Drying Characteristics of Rubberwood by Impinging Hot-Air and Microwave Heating." *Wood Research* 60 (1): 59–70.

Rufino Franco, C. M., and A. G. Barbosa de Lima. 2016. "Intermittent Drying of Porous Materials: A Review." *Diffusion Foundations*, Diffus. Found., 7: 1–17. doi:10.4028/www.scientific.net/DF.7.1.

Sailor, David J., Daniel J. Rohli, and Qianli Fu. 1999. "Effect of Variable Duty Cycle Flow Pulsations on Heat Transfer Enhancement for an Impinging Air Jet." *International Journal of Heat and Fluid Flow* 20 (6): 574–80. doi:10.1016/S0142-727X(99)00055-7.

Sexton, Andrew, Jeff Punch, Jason Stafford, and Nicholas Jeffers. 2018. "The Thermal and Hydrodynamic Behaviour of Confined, Normally Impinging Laminar Slot Jets." *International Journal of Heat and Mass Transfer* 123 (August): 40–53. doi:10.1016/j. ijheatmasstransfer.2018.02.083.

Sheriff, H. S., and D. A. Zumbrunnen. 1994. "Effect of Flow Pulsations on the Cooling Effectiveness of an Impinging Jet." *Journal of Heat Transfer* 116 (4): 886–95. doi:10.1115/1.2911463.

Silva, Wilton Pereira da, Andréa Fernandes Rodrigues, Cleide Maria D. P. S. e Silva, Deise Souza de Castro, and Josivanda Palmeira Gomes. 2015. "Comparison between Continuous and Intermittent Drying of Whole Bananas Using Empirical and Diffusion Models to Describe the Processes." *Journal of Food Engineering* 166 (December): 230–36. doi:10.1016/j.jfoodeng.2015.06.018.

Silva, V., A. R. Figueiredo, J. J. Costa, and R. P. F. Guiné. 2014. "Experimental and Mathematical Study of the Discontinuous Drying Kinetics of Pears." *Journal of Food Engineering* 134 (August): 30–36. doi:10.1016/j.jfoodeng.2014.02.022.

Silva, Vítor, José J. Costa, A. Rui Figueiredo, João Nunes, Catarina Nunes, Tânia I. B. Ribeiro, and Bruno Pereira. 2016. "Study of Three-Stage Intermittent Drying of Pears Considering Shrinkage and Variable Diffusion Coefficient." *Journal of Food Engineering* 180 (July): 77–86. doi:10.1016/j.jfoodeng.2016.02.013.

Specht, Eckehard. 2014. "Impinging Jet Drying." In *Modern Drying Technology*, 1–26. Wiley-Blackwell. doi:10.1002/9783527631704.ch01.

Tapaneyasin, Rungtip, Sakamon Devahastin, and Ampawan Tansakul. 2005. "Drying Methods and Quality of Shrimp Dried in a Jet-Spouted Bed Dryer." *Journal of Food Process Engineering* 28 (1): 35–52. doi:10.1111/j.1745-4530.2005.00394.x.

Tawfek, A. A. 1996. "Heat Transfer and Pressure Distributions of an Impinging Jet on a Flat Surface." *Heat and Mass Transfer* 32 (1–2): 49–54. doi:10.1007/s002310050090.

Thurlow, G. G. 1954. "Communications on Heat Transfer by Convection." *Proceedings of the Institution of Mechanical Engineers* 168 (30): 781–83.

Wachiraphansakul, Sarat, and Sakamon Devahastin. 2005. "Drying Kinetics and Quality of Soy Residue (Okara) Dried in a Jet Spouted-Bed Dryer." *Drying Technology* 23 (6): 1229–42. doi:10.1081/DRT-200059421.

Wåhlby, Urban, Christina Skjöldebrand, and Elmar Junker. 2000. "Impact of Impingement on Cooking Time and Food Quality." *Journal of Food Engineering* 43 (3): 179–87. doi:10.1016/S0260-8774(99)00149-1.

Wen, Mao-Yu, and Kuen-Jang Jang. 2003. "An Impingement Cooling on a Flat Surface by Using Circular Jet with Longitudinal Swirling Strips." *International Journal of Heat and Mass Transfer* 46 (24): 4657–67. doi:10.1016/S0017-9310(03)00302-8.

Wu, Zhong Hua, Arun S. Mujumdar, X.D. Liu, and L. Yang. 2006. "Pulse Combustion Impingement to Enhance Paper Drying." In *Proceedings of the 15th International Drying Symposium*, edited by István Farkas. Budapest, Hungary.

Xiao, Hong-Wei, Jun-Wen Bai, Long Xie, Da-Wen Sun, and Zhen-Jiang Gao. 2015. "Thin-Layer Air Impingement Drying Enhances Drying Rate of American Ginseng (Panax Quinquefolium L.) Slices with Quality Attributes Considered." *Food and Bioproducts Processing* 94 (April): 581–91. doi:10.1016/j.fbp.2014.08.008.

Xiao, Hong-Wei, Zhen-Jiang Gao, Hai Lin, and Wen-Xia Yang. 2010a. "Air Impingement Drying Characteristics and Quality of Carrot Cubes." *Journal of Food Process Engineering* 33 (5): 899–918. doi:10.1111/j.1745-4530.2008.00314.x.

Xiao, Hong-Wei, Chung-Lim Law, Da-Wen Sun, and Zhen-Jiang Gao. 2014. "Color Change Kinetics of American Ginseng (Panax Quinquefolium) Slices During Air Impingement Drying." *Drying Technology* 32 (4): 418–27. doi:10.1080/07373937.2013. 834928.

Xiao, Hong-Wei, and A. Mujumdar. 2014. "Impingement Drying: Application and Future Trends." In Prabhat K. Nema, Barjinder Pal Kaur, Arun S. Mujumdar, Drying Technologies for Foods: Fundamentals & Applications (ISBN978-938-14-5074-1) New India Publishing Agency, New Delhi, India (pp. 279–299).

Xiao, Hong-Wei, Chang-Le Pang, Li-Hong Wang, Jun-Wen Bai, Wen-Xia Yang, and Zhen-Jiang Gao. 2010b. "Drying Kinetics and Quality of Monukka Seedless Grapes Dried in an Air-Impingement Jet Dryer." *Biosystems Engineering* 105 (2): 233–40. doi:10.1016/j.biosystemseng.2009.11.001.

Xu, Peng, Shuxia Qiu, MingZhou Yu, Xianwu Qiao, and Arun S. Mujumdar. 2012. "A Study on the Heat and Mass Transfer Properties of Multiple Pulsating Impinging Jets." *International Communications in Heat and Mass Transfer* 39 (3): 378–82. doi:10.1016/j.icheatmasstransfer.2012.01.001.

Yamsaengsung, Ram, and Oraporn Bualuang. 2010. "Air Impingement Drying of Spirulina Platensis." *Songklanakarin Journal of Science and Technology* 32 (1): 52–62.

Zargarabadi, Mehran Rajabi, Ehsan Rezaei, and Babak Yousefi-Lafouraki. 2018. "Numerical Analysis of Turbulent Flow and Heat Transfer of Sinusoidal Pulsed Jet Impinging on an Asymmetrical Concave Surface." *Applied Thermal Engineering* 128 (January): 578–85. doi:10.1016/j.applthermaleng.2017.09.059.

Zhang, Qian, Hongwei Xiao, Jianwu Dai, Xuhai Yang, Junwen Bai, Zheng Lou, and Zhenjiang Gao. 2011. "Air Impingement Drying Characteristics and Drying Model of Hami Melon Flake." *Nongye Gongcheng Xuebao/Transactions of the Chinese Society of Agricultural Engineering* 27 (1): 382–88.

Zhang, Yanyao, Ping Li, and Yonghui Xie. 2018. "Numerical Investigation of Heat Transfer Characteristics of Impinging Synthetic Jets with Different Waveforms." *International Journal of Heat and Mass Transfer* 125 (October): 1017–27. doi:10.1016/j.ijheatmasstransfer.2018.04.120.

Zheng, Xia, Hongwei Xiao, Lihong Wang, Qian Zhang, Junwen Bai, Long Xie, Haoyu Ju, and Zhenjiang Gao. 2014. "Shorting Drying Time of Hami-Melon Slice Using Infrared Radiation Combined with Air Impingement Drying." Text. January 1. http://www.ingentaconnect.com/content/tcsae/tcsae/2014/00000030/00000001/art00033#.

Zumbrunnen, D. A., and M. Balasubramanian. 1995. "Convective Heat Transfer Enhancement Due to Gas Injection Into an Impinging Liquid Jet." *Journal of Heat Transfer* 117 (4): 1011–17. doi:10.1115/1.2836275.

6 Wicking and Drying in Fibrous Porous Materials

Dahua Shou
The Hong Kong Polytechnic University,
Hung Hom, Kowloon, Hong Kong

Jintu Fan
Cornell University, Ithaca, New York, United States

CONTENTS

6.1 INTRODUCTION TO DRYING IN FIBROUS POROUS MATERIALS

Fibrous porous materials are generally solid-void mixtures with the solid fibers in a slender circular/elliptical form (see Figure 6.1). The unique properties of fibrous materials, different from some other categories of porous media, are their high specific surface area, breathability, compressibility, and flexibility. Particularly in textile engineering, fibrous materials can be mainly divided into two types: ordered fabrics and disordered nonwovens.

Fibrous porous materials in nonwoven form are generated by means of assembling fibers into a planar pattern. Although the nature of their geometrical structures are complex, the nonwovens can be generally simplified as consisting of bundles of cylindrical fibers in three forms, viz., one-dimensional (1D) structure in which all fibers are parallel with each other; two-dimensional (2D) structure in which fibers are placed in parallel planes with directional or random orientations;

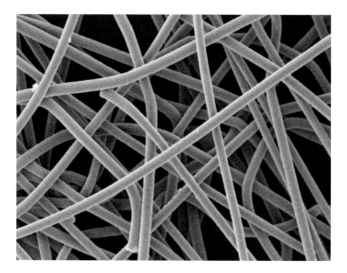

FIGURE 6.1   A microscopic image of a fibrous porous medium.

and three-dimensional (3D) structure in which fibers are directionally or randomly oriented in space (Tomadakis and Robertson 2005). Ordered fabrics are made up of bundles of fibers after arrangements, such as weaving, knitting, crocheting, and knotting. Among them, woven structures are the ones that are most widely applied in fiber-based composites (Chen, Ye, and Kruckenberg 2006), which have additional features including unit cell, interlace spacing or gap, and interlace point (Mariatti, Nasir, and Ismail 2000). The popularity of woven composites is increasing due to the dual-scale structures. Woven fabric is made up of filament bundles, known as yarns. The weaving of the yarns provides an additional interlocking which enhances strength better than what can be achieved by sole fiber matrix adhesion (Alavudeen et al. 2011).

The fibrous material is always composed of interconnected cavities, grooves, or hollow channels of disordered size and shape. When the fibrous system is wet, these porous spaces will be absorbed and filled with different degrees of liquids, such as water, due to capillary action. The spaces within the fibrous construct contains water vapor, liquid water, and air. With the temperature difference or gradient, drying occurs as a mass transport process that consists of the transfer of water or another liquid by evaporation from the fiber surface.

In the drying process, the most widely used heating modes that utilize thermal energy transfer from a heat source to the wetted fibers include radiation, conduction, convection, or a combination of these. No matter how the fiber body is heated or where the vaporization site is located, the liquid must be transported by flow means, such as wicking from the liquid source to the site of vaporization. Radiation heating is a method for rapid thermal energy transfer to the fibrous material without physical contact, on the basis of absorption of infrared radiation emitted from radiators such as the sun with higher temperature. It is noted that water has relatively high absorption coefficient of thermal radiation. If the fiber surface is saturated or covered with a layer of water film, the surface temperature will often not be over the boiling point of water for the evaporative cooling effect. Conduction heating refers to less heat resistance between the fibrous systems and the heat source, when the fibers are in physical contact with a heated spot or surface. Therefore, the heat transfer rate is considerably high when the fibers are composed of thermally conductive materials such as metallic wires. However, over heating may cause damage to the fibrous materials in the contacting area when the drying occurs extremely quickly. Convection heat is brought by the ventilation of dry and heated air over the fibrous surface. A thermal boundary layer on the fiber surface is often introduced as a thermal resistance to reduce the drying or evaporation processes.

There are several types of moisture transport mechanisms within the fibrous systems, which mainly include wicking and diffusion driven by capillary pressure and concentration differences, respectively, and the vaporization-condensation sequence. There are also a few important thermal energy transport mechanisms, such as thermal conduction through the solid fibers and void pores, convective heat transfers by the moving fluids, thermal radiation within the fibrous structure due to temperature difference, and the transfer of latent heat by vaporization. In particular, the coupling between the thermal energy and moisture transport mechanisms should be simultaneously analyzed in the drying process.

6.2 BASIC HEAT AND MOISTURE TRANSFER PROCESS IN DRYING OF FIBROUS POROUS MATERIALS

Drying on the basis of coupled heat and moisture transfer in fibrous porous materials started in the 1930s. However, little further progress had been made until the 1980s, when the pioneering researchers explored this subject by theoretical modeling and numerical simulation, assuming that heat is transported by conduction and convection (Ogniewicz and Tien 1981). Later, the quasi-steady state corresponding to mobile liquid in the fibrous systems was investigated, where the liquid diffuses in the wet zone's boundaries with consideration of evaporation at these boundaries (Motakef and Elmasri 1986). More comprehensive models of heat and moisture transfer for drying process in fibrous porous materials have been reported and analyzed, illustrating the heat and moisture transfer process dynamically (Fan et al. 2004).

The typical model considers a thick fibrous porous batting (sandwiched by two thin fabrics) next to the human skin, which is the heat and liquid source, and with the environment on its other side, with a lower temperature (Figure 6.2). Radiative heat transfer is considered because of the large temperature difference between the skin side and the external environment.

Based on the conservation of thermal energy at position x and time t, the heat transfer equation can be described by

$$C_v(x,t)\frac{\partial T}{\partial t} = -\varepsilon C_{vv}(x,t)\frac{\partial T}{\partial x} + \frac{\partial}{\partial t}\left[k(x,t)\frac{\partial T}{\partial x}\right] + \frac{\partial F_R}{\partial x} - \frac{\partial F_L}{\partial x} + \lambda(x,t)\Gamma(x,t), \quad (6.1)$$

where $k(x,t)$ is the effective thermal conductivity, $C_v(x,t)$ is the effective volumetric heat capacity of the fibrous batting, $C_{vv}(x,t)$ the effective volumetric heat capacity

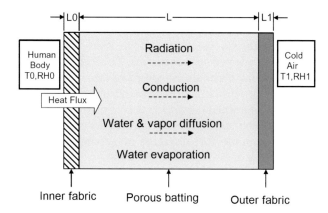

FIGURE 6.2 Schematic diagram of heat and moisture transfer for drying in the fibrous porous ensemble. (Reprinted with permission from J. Fan and X. Wen, "Modeling heat and moisture transfer through fibrous insulation with phase change and mobile condensates." *International Journal of Heat and Mass Transfer* 45 (2002): 4045–4055.)

of the water vapor, ε is porosity of the fibrous medium, T is temperature, $\lambda(x, t)$ is latent heat of water, $\Gamma(x, t)$ is the rate of evaporation or condensation, F_R is thermal radiation incident at point xx traveling to the right, and F_L is the total thermal radiation incident traveling to the left. The radiative heat absorbed by the fibers can be characterized by an absorption constant β, and the thermal emissivity of the volume element is βdx. Thus the attenuation of the radiation fluxes can be described as follows,

$$\frac{\partial F_R}{\partial x} = -\beta F_R + \beta\sigma T^4(x, t),\tag{6.2}$$

$$\frac{\partial F_L}{\partial x} = \beta F_L - \beta\sigma T^4(x, t).\tag{6.3}$$

where σ is the Boltzmann constant. Based on mass conservation, the water vapor transfer process for drying in the fibrous system is characterized by the moisture transfer equation:

$$\varepsilon\frac{\partial C_a}{\partial t} = -\varepsilon u\frac{\partial C_a}{\partial x} + \frac{D_a\varepsilon}{\tau}\frac{\partial^2 C_a}{\partial x^2} - \Gamma(x, t),\tag{6.4}$$

where D_a is the diffusion coefficient of water vapor in the air and u is the flow velocity of water vapor. The vapor diffusion often follows Fick's law within the fibrous porous medium. Similarly, the dynamics of the water transport is described as the following on the basis of mass conservation,

$$\rho(1-\varepsilon)\frac{\partial W}{\partial t} = \rho(1-\varepsilon)D_l\frac{\partial^2 W}{\partial x^2} + \Gamma_{ce}(x, t),\tag{6.5}$$

where D_l is the diffusion coefficient of liquid water, W is the water content of the fibrous layer, and $\Gamma_{ce}(x, t)$ is the rate of condensation or evaporation.

The significant mechanism directly referring to drying is the evaporation and condensation process, which is widely described by the Hertz-Knudsen equation,

$$\Gamma_{ce}\left(x,\,t\right) = -E\sqrt{M\,/\,2\pi R_u}\left(\frac{P_{sat}}{\sqrt{T_s}} - \frac{P_v}{\sqrt{T_v}}\right),\qquad(6.6)$$

where E is the evaporation or condensation coefficient, M is the molecular weight of water, R_u is the universal gas constant, P_{sat} is the saturated water vapor pressure, P_v is the water vapor pressure in the air, T_s is the temperature at the interface between vapor and liquid, and T_v is the water vapor temperature in the air. With proper assumptions and with boundary conditions constrained by the practical problems, the dynamic drying process can be obtained by solving the above equations.

6.3 BASIC WICKING AND DRYING PROCESS

As a passive mode of liquid transport, capillary-driven wicking exists in numerous applications including microfluidic diagnostics, microscale molding and manufacturing, functional fabrics, and oil–water separation. Wicking-based drying or evaporative cooling has received increasing attention in solar steam generation (Ghasemi et al. 2014), microscale heat exchanger (Cai and Bhunia 2012), personal thermal management (Fan and Wen 2002), and building thermal comfort (Chen, Liu, and Lin 2015), considering the efficient heat transfer by phase change. Moreover, in today's information age, cooling in particular has become a major challenge in data centers, accounting for roughly 30–40% of their total energy use. Cooling technology is the key to the reducing the energy consumption of heating, ventilation, and air conditioning systems, accounting for about 13% of the total energy consumed in the U.S. and about 40% of the energy used in a typical U.S. residence, making it the largest energy expense for most homes.

Nanoscale and microscale devices and systems are increasingly used to enhance capillary force and surface area in favor of flow and evaporative drying, respectively (Cai et al. 2014, Bergles et al. 2003, Adera et al. 2016). However, fluid speed and evaporation area are often bottlenecks in advancing these functionalities, due to increased flow resistance (Shou and Fan 2018, Liu et al. 2016). For example, floating capillary systems incorporating nanomaterials and metallic structures were recently developed to wick water and localize the heat at the top face of the bulk water for steam generation (Ni et al. 2016). Thin cellulose-layered cloths with a single porous rod were introduced to form a water path to minimize heat loss of bulk water, with energy conversion efficiency approaching 94% (Li et al. 2016, Xu et al. 2017), but the optimization of steam generators for faster capillary flow and enhanced drying has been less investigated, especially considering the effect of gravity. Furthermore, thin-film evaporation in nano- and microporous systems has shown ultra-efficient cooling ability for thermoregulation of power microelectronics (Cai and Bhunia 2012), integrated circuits, and laser diodes, in which the heat flux can be over several hundred watts per square centimeter (Xiao, Maroo, and Wang 2013). However, implementing thin-film evaporation by delivering liquid to a large region in sufficient quantity to feed the evaporation remains a significant challenge in tiny porous systems. Moreover, latent heat transfer on the basis of wicking and drying is a classic topic in the field of thermal comfort and clothing physiology (Fan, Luo, and Li 2000). Various fibrous materials have been experimentally studied to demonstrate their performance for indirect evaporative cooling applications (Xu et al. 2016), but optimized guidelines to enhance the latent heat transfer by drying and reduce the weight of the fabrics is desirable.

Much effort has been directed to the modeling and regulation of wicking-driven drying in different materials and structures, considering the significant promise in passive applications without energy consumption. For the simplest representative structure of the uniform hollow tube, the capillary flow follows a diffusive law (Washburn 1921), $h = \sqrt{Dt}$, where D is the diffusive coefficient determined by the pore size, surface tension, and contact angle. For a porous system, the flow behavior can also be described by the diffusive relationship with negligible effects of drying and gravity, where the interconnected pores are approximated as bundles of tortuous channels (Cai et al. 2012). In a non-uniform, conical tube with axial variation in the radius, a different power law $h \sim t^{1/4}$ has been found for long durations (Reyssat et al. 2008). Furthermore, the power law becomes $h \sim t^{1/3}$ for the radial liquid absorption in a semi-infinite porous medium (Xiao, Stone, and Attinger 2012). Thus, the structural design provides a potential tool for regulating or tuning capillary flow velocity. In particular, wicking is accelerated through enhancement of the effective pore size of porous systems by packing two layers of chromatography paper with a hollow gap between them (Camplisson et al. 2015), carving small open grooves on the face of chromatography paper (Giokas, Tsogas, and Vlessidis 2014), and sandwiching a filter paper between flexible plastic films (Jahanshahi-Anbuhi et al. 2012). Recently, the fastest wicking was found from multi-section tubes to multi-layer porous systems with and without the gravitational role, based on an optimized trade-off between viscous resistance, capillary drag, and gravitational force (Shou, Ye, and Fan 2014b, Shou, Ye, et al. 2014, Shou and Fan 2015a). It is interesting to note that the flow velocity can be tuned even at a fixed value in an optimized porous layer with varied thickness or width (Shou and Fan 2015b). When gravity is in action under realistic conditions, capillary rise is slowed significantly in the long run due to the increasing gravitational force (Figure 6.3), which makes flow enhancement more challenging (Shou, Ye, and Fan 2014a). The capillary flow dynamics become markedly impacted by the gravitational and evaporative effects with increasing time. Wicking behaviors limited by drying have also been characterized at different drying rates, whereas the geometrical effects are found to enable varying liquid wicking behaviors in porous structures (Liu et al. 2016).

Fast and controllable liquid delivery is desirable for all the above mentioned applications and the optimal structural parameters remain to be determined under evaporative drying and gravitational force. It is also critical to identify structures with maximum evaporation rate or drying area. The traditional approach to increasing width or thickness will increase the thermal resistance of the total system. It is critical to enhance the drying effect but without compromising the thermal transfer property. To this end, enhancing and programming the wicking and drying processes for advanced applications are demonstrated by varying the structural parameters of the porous fibrous systems.

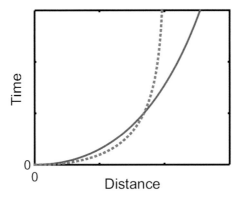

FIGURE 6.3 Comparison of capillary rise in two uniform tubes with different sizes under gravity. The solid line refers to a tube with a smaller radius and the dotted line refers to a tube with a greater radius. The capillary rise is faster in the coarser tube in the beginning stage, but it becomes slower and then easier to suspend in the long run, based on the Lucas-Washburn equation.

6.4 MODELS OF WICKING AND DRYING

6.4.1 WICKING IN A TUBE

When a liquid encounters a hollow cylindrical tube, a rise of the liquid occurs in the tube. This process is known as capillary flow or wicking, by which a liquid moves in a narrow space without external assistance such as gravitational force. During the wicking process, drying always occurs as a mass transfer process with loss of water or another liquid by evaporation. The momentum balance for the wicking process is given by

$$p_c = \Delta p + p_g, \tag{6.7}$$

where the capillary pressure p_c equals to the sums of pressures corresponding to viscous friction Δp and gravity p_g, respectively. Here, the capillary pressure jumping across the liquid front surface is related to the tube radius R, the surface tension γ and the contact angle θ,

$$p_c = -\frac{2\gamma cos\theta}{R}. \tag{6.8}$$

The hydrostatic pressure due to the gravitational force is widely expressed as

$$p_g = -\rho g h sin\alpha, \tag{6.9}$$

where g is the gravity, ρ is the fluid density, h is the flow distance of the fluid front, and α is the incline angle of the tubing system. The pressure due to the viscous pressure is described by the Hagen–Poiseuille equation, viz.,

$$p_c = -\frac{8\mu}{R^2}\frac{dh}{dt}h, \tag{6.10}$$

where μ is the viscosity of the liquid. When the gravity is negligible for a horizontally placed tube or in the early stage of wicking, the Lucas-Washburn equation can be obtained by the initial condition $h(t=0)=0$,

$$t = 2C\frac{h^2}{R},\tag{6.11}$$

where C is a constant given by $\frac{\mu}{\gamma cos\theta}$. It is interesting to note that the square of the flow distance is linear with the time without the effects of gravitational or inertia forces. A maximum distance or height can be obtained when the fluid front stops rising at $\frac{dh}{dt}=0$, viz.,

$$h_{max} = \frac{2\gamma cos\theta}{\rho gRsin\alpha}.\tag{6.12}$$

6.4.2 WICKING AND DRYING IN A FIBROUS MEDIUM

For fibrous porous systems, which can always be assumed to consist of bundles of tortuous tubes (Figure 6.4), the pressure due to the viscous friction is given by

$$\Delta p = -\frac{\mu\varepsilon_s h}{K_p}\frac{dh}{dt},\tag{6.13}$$

where K_p is the hydraulic permeability and ε_s is the porosity of the porous strip. Here, the hydraulic permeability of the porous system is expressed as

$$K_p = \frac{R^2\varepsilon_s}{8\tau},\tag{6.14}$$

where τ is the tortuosity factor corresponding to the tortuous paths in the porous media and R is the mean tube radius. Here the tortuosity is often approximated to be a function of porosity, viz., $\tau = \varepsilon^{-0.5}$. The permeability in Eq. (6.14) can be replaced by accurately predicted models of the fiber orientation and construction arrangements.

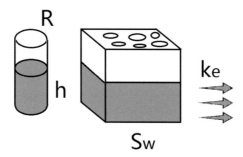

FIGURE 6.4 A representative hollow tube for wicking (left) and a fibrous porous medium for wicking and evaporative drying (right).

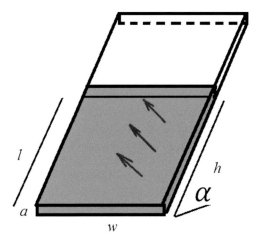

FIGURE 6.5   Schematic illustration of wicking and drying in an inclined fibrous medium.

Substituting Eq. (6.13) and Eq. (6.14) into Eq. (6.7) gives the model of the dynamical flow process,

$$\frac{dh}{dt} = \frac{\gamma R cos\theta}{4\mu\tau}\frac{1}{h} - \frac{\rho g R^2 sin\alpha}{8\mu\tau}. \tag{6.15}$$

The drying caused by evaporation loss of liquid (Figure 6.5) becomes significant in the long term or at high temperature, and the loss of mass is generally proportional to the area of the wetted surface exposed to ambient air. Based on the law of conservation of mass, the flow rate at h is the sum of that at l and the flow rate due to the evaporation loss,

$$Q(l) = Q(h) + k_e S_w, \tag{6.16}$$

where k_e is the evaporation rate per area, $S_w = w(l-h)$ is area of the wetted surface between h and l, and w is the width of the fibrous medium. The side area of a thin porous layer is much smaller than the face area, so the drying on the side surface can be ignored. $Q(l)$ and $Q(h)$ are the effective flow rates at l and h, respectively, viz.,

$$Q(l) = \frac{dl}{dt}aw\varepsilon_s \tag{6.17}$$

and

$$Q(h) = \frac{dh}{dt}aw\varepsilon_s, \tag{6.18}$$

where $aw\varepsilon$ represents the void area of the cross section in the porous layer and a is the thickness of the fibrous medium. Substituting Eq. (6.17) and Eq. (6.18) into Eq. (6.16) gives

$$\frac{dl}{dt} = \frac{dh}{dt} + \frac{k_e(l-h)}{a\varepsilon}. \tag{6.19}$$

The pressure difference corresponding the combined effects of viscous friction and drying between 0 and h is given by

$$\Delta p = -\int_0^h \frac{\mu \varepsilon_s}{K_p \tau} \left(\frac{dh}{dt} + \frac{k_e (l-h)}{a\varepsilon} \right) dl = -\frac{\mu \varepsilon_s h}{K_p \tau} \frac{dh}{dt} - \frac{1}{2} \frac{\mu k_e h^2}{a K_p \tau} \qquad (6.20)$$

Hence, the flow velocity of the liquid is modified considering the drying effect, viz.,

$$\frac{dh}{dt} = \frac{\gamma R \cos\theta}{4\mu\tau} \frac{1}{h} - \frac{\rho g R^2 \sin\alpha}{8\mu\tau} - \frac{k_e h}{2l_t \rho \varepsilon_s}. \qquad (6.21)$$

Eq. (6.21) can be rewritten in a simplified form:

$$\frac{dh}{dt} = A_1 \frac{1}{h} - B_1 - C_1 h, \qquad (6.22)$$

with the constants defined as $A_1 = \frac{\gamma R \cos\theta}{4\mu\varepsilon_s^{-0.5}}$, $B_1 = \frac{\rho g R^2 \sin\alpha}{8\mu\varepsilon_s^{-0.5}}$, and $C_1 = \frac{k_e}{2l_t \rho \varepsilon_s}$.

When the effects of drying and gravity are negligible, Eq. (6.22) is reduced to the form:

$$\frac{dh}{dt} = A_1 \frac{1}{h}, \qquad (6.23)$$

Thus the Lucas-Washburn relationship for the porous fibrous system is obtained, viz.,

$$t = \frac{h^2}{2A_1}. \qquad (6.24)$$

In addition, the maximum final height or spreading distance of the liquid in the fibrous system can be determined by calculating Eq. (6.22) with $\frac{dh}{dt} = 0$, when the effects of the evaporation loss and gravitational force are practically significant.

6.4.3 Optimized Wicking and Drying in a Fibrous Medium

For many drying-based applications it is expected that the drying area is maximized for evaporation while the liquid delivery is sufficiently fast. It is known that an increase in pore size will lead to decreases in capillary pressure and viscous resistance and enhancement of the gravitational force. As well, the reduction of flow velocity caused by the gravitational force and evaporation loss will be more significant in the long term. Thus it is hypothesized that the fastest capillary flow can be found against the pore size for maximizing drying or evaporation in fibrous materials.

By integrating Eq. (6.22) the flow time for the wicking distance is obtained at different drying rates. Note that $k_e = 7 \times 10^{-6}$ kg / m^{2}s corresponds to the evaporation rate of water from a water surface without wind at room temperature. Here, the thickness of the fibrous media is $a = 0.001$ m, the porosity is $\varepsilon = 0.5$, the incline angle is $\alpha = \pi / 2$, the contact angle is $\theta = 0$, and the given flow distance is $h_0 = 0.1$ m.

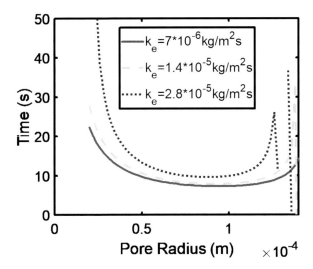

FIGURE 6.6   Flow time versus pore radii in a fibrous medium at different drying rates. Here, the thickness of the fibrous mediums is $a = 0.001$ m, the porosity is $\varepsilon = 0.5$, the incline angle is $\alpha = \pi / 2$, the contact angle is $\theta = 0$, and the given flow distance is $h_0 = 0.1$ m.

It is evident that the minimum time accounting for the fastest flow is found, validating that the pore size can be optimized for maximum wicking and drying (Figure 6.6). It is also noted that with increasing drying rates, the flow time increases or the flow velocity reduces at the same pore size. A sudden decrease in time is found at around $R = 13$ μm for the drying rate at $k_e = 2.8 \times 10^{-5}$ kg / m^{2}s, which indicates that the maximum height has been found. The model fails or the liquid stops advancing when the pore sizes exceeds $R = 13$ μm. It is also demonstrated that the flow time increases dramatically when the pore size is less than $R = 50$ μm, considering the dramatically enhanced flow resistance.

For a given flow distance at $h_0 = 0.15$ m, the flow time increases considerably. It is interesting to note that the optimized pore size increases to the range close to $R = 60$ μm (Figure 6.7), in comparison to cases with the given flow distance at $h_0 = 0.1$ m. A sudden reduction in flow time is found when the pore radius is less than $R = 34$ μm for the drying rate at $k_e = 1.4 \times 10^{-5}$ kg / m^{2}s, an effect that is attributed to drying-up or to insufficient liquid delivery in the smaller pore regimes with a higher flow resistance. It is also apparent that the range of pore sizes for liquid delivery becomes much narrower at the drying rate $k_e = 1.4 \times 10^{-5}$ kg / m^{2}s. Therefore, it is critical to select the proper pore sizes for maintaining high wicking and drying.

6.4.4   WICKING AND DRYING IN LAYERED FIBROUS MATERIALS

To further enhance the wicking in fibrous porous materials, multi-layer fibrous channels are developed to accelerate the liquid delivery. Inspired by the microstructure of the wedged cavities on the surface of *Nepenthes alata*'s peristome, which allows fast

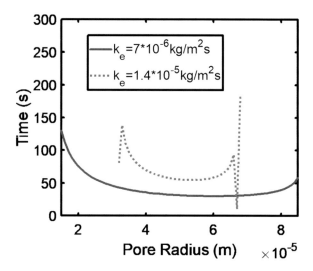

FIGURE 6.7 Flow time versus pore radii in a fibrous medium at different drying rates. Here, the thickness of fibrous media is $a = 0.001$ m, the porosity is $\varepsilon = 0.5$, the incline angle is $\alpha = \pi / 2$, the contact angle is $\theta = 0$, the given flow distance is $h_0 = 0.15$ m.

and continuous directional water transport (Chen et al. 2016), an innovative fibrous device has been proposed to accelerate wicking during drying. The device is a hollow wedged channel (W-Channel) made from nano-/microfibrous layers, as seen in Figure 6.8. The wicking and drying are enhanced through an adaptive interplay of capillary pressure, evaporation loss, gravitational force, and viscous friction. The

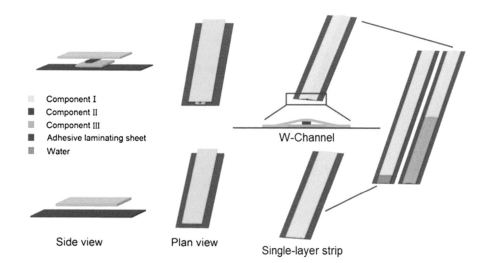

FIGURE 6.8 Schematic illustration of the fabrication process of W-Channels and single-layer strips. (Reprinted with permission from (Shou and Fan 2018). Copyright (2018) American Chemical Society).

development of the W-Channel is shown by theoretical modeling in this section, on which basis a device fabricated using both microfibrous filter papers and electrospun nanofibers is evaluated experimentally in the next section.

On the basis of Darcy's law, the pressure drop of liquid in the triangular hollow wedge between fibrous layers due to the viscous friction is given by (Figure 6.9)

$$\Delta p = -\frac{\mu h}{K_t}\frac{dh}{dt},\tag{6.25}$$

where the permeability of the channel scales with the square of the inscribed radius of the wedge channel r_{in}, viz., $K_t = \frac{3}{80}r_{in}^2$ (Jia et al. 2008).

On the basis of the minimum free energy approach (Butt, Graf, and Kappl 2006), the change in free energy dU upon an infinitesimal disturbance dh of the fluid front from the equilibrium position can be expressed as

$$dU = \gamma_{SL}dA_{SL} + \gamma_{SV}dA_{SV} + \gamma_{VL}dA_{VL},\tag{6.26}$$

where γ_{SL}, γ_{SV}, and γ_{VL} are the solid–liquid, solid–vapor, and vapor–liquid interfacial tensions, respectively, and A_{SL}, A_{SV}, and A_{VL} are the solid–liquid, solid–vapor, and vapor–liquid interfacial areas, respectively.

The wicking fluid in the surrounding fibrous layers is coupled with that in the hollow triangular wedge. Generally, the flow velocity in the fibrous layers with much smaller pores is slower than in the hollow channel. However, the initial liquid absorption by the porous strip is very fast, considering the extremely small thickness of the strip. Therefore, the flow velocities of the liquid in the porous layers and the hollow triangular wedges are approximately similar. The modified flow velocity considering the water absorption by the porous fibrous layers with the correction factor $\frac{S_{tri}}{S_{tri}+\varepsilon S_{strip}}$ is expressed as

$$\frac{dh}{dt} = \frac{S_{tri}}{S_{tri}+\varepsilon S_{strip}}\left\{\frac{3r_{in}^2\left(1+tan\omega+1/cos\omega\right)\gamma cos\theta}{40\left[\dfrac{r_{in}}{tan(\omega/2)}+r_{in}\right]\mu htan\omega}+\frac{3\rho gsin\alpha}{80}r_{in}^2\right\},\tag{6.27}$$

where S_{strip} and ε are the total cross-sectional area and porosity of the porous strips, respectively. ε becomes equal to ε_s when components I, II, and III are made from the same materials.

Analogous to the case of a single-layer fibrous medium, Eq. (6.27) is modified by taking into account the evaporative drying effect and a correction term $-\frac{k_e wh}{\rho\left(S_{tri}+\varepsilon S_{strip}\right)}$ is added to reduce the flow velocity accordingly,

$$\frac{dh}{dt} = \frac{S_{tri}}{S_{tri}+\varepsilon S_{strip}}\left\{\frac{3r_{in}^2\left(1+tan\omega+\dfrac{1}{cos\omega}\right)\gamma cos\theta}{40\left[\dfrac{r_{in}}{tan(\omega/2)}+r_{in}\right]\mu htan\omega}+\frac{3\rho gsin\alpha}{80}r_{in}^2\right\}-\frac{k_e wh}{\rho\left(S_{tri}+\varepsilon S_{strip}\right)}.$$

$$\tag{6.28}$$

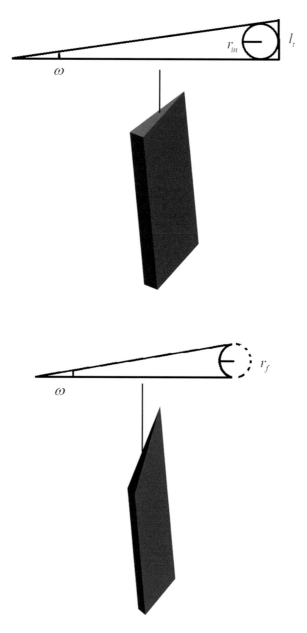

FIGURE 6.9 Illustration of capillary rise in W-Channels (a) when the front moves along the closed wedge in Stage 1 and (b) when the front approaches the edge of the wedge in Stage 2, where ω is the angle between the two flat fibrous layers, r_{in} is the inscribed radius of the triangular wedge channel, l_t is the thickness of the fibrous layer, and r_f is the radius of the curvature of the fluid front.

The transport period characterized by Eq. (6.28) is called Stage 1. Eq. (6.28) can also be rewritten in a simplified form,

$$\frac{dh}{dt} = A_2 \frac{1}{h} - B_2 - C_2 h, \tag{6.29}$$

where the constants are given by $A_1 = \dfrac{3r_{in}^2(1+\tan\omega+1/\cos\omega)\gamma\cos\theta}{40\left[\dfrac{r_{in}}{\tan(\omega/2)}+r_{in}\right]\mu\tan\omega} \dfrac{S_{tri}}{S_{tri}+\varepsilon S_{strip}}$, $B_1 = \dfrac{3\rho g\sin\alpha}{80} \dfrac{S_{tri}}{S_{tri}+\varepsilon S_{strip}} r_{in}^2$,

and $C_1 = \dfrac{k_e w}{\rho\left(S_{tri}+\varepsilon S_{strip}\right)}$, respectively.

With the increasing rise of the fluid in the wedge, the front stops progressing vertically in the wider area because the increasing gravitational pressure is equal to the capillary pressure. But the fluid front continues to advance in the narrow area with decreasing spacing and approaches the corner edge of the wedge. Thus the cross-sectional area of the fluid front surface decreases and the transport process in this period is called Stage 2.

Then the capillary pressure varies with the rise of the fluid front, which is balanced by the pressures caused by gravitational force and viscous friction. The scaling law for the capillary rise between the two flat plates in the wedge is analogous to that in the corner of two closely packed cylinders (Ponomarenko, Quere, and Clanet 2011), viz.,

$$p_{c,f} \sim \rho gh\sin\alpha - \mu\frac{h}{r_f^2}\frac{dh}{dt}, \tag{6.30}$$

where r_f is the radius of the curvature of the fluid front with the height of h (Figure 6.9). Here, h also indicates the location of the fluid front in the tube of radius r_f at time t.

When the evaporation loss for drying is considered, Eq. (6.29) can be modified as

$$\frac{dh}{dt} \sim \frac{S_{tri,m}}{S_{tri,m}+\varepsilon S_{strip}}\left\{\frac{\gamma\left[2\cos\theta+(\pi-\omega)\tan(\omega/2)\right]}{\left(1-\dfrac{\pi-\omega}{2}\tan(\omega/2)\right)\mu h}r_f - r_f^2\frac{\rho g\sin\alpha}{\mu}\right\}$$

$$- \frac{k_e wh}{\rho\left(S_{tri,m}+\varepsilon S_{strip}\right)}. \tag{6.31}$$

When the drying is less significant in the early stage, Eq. (6.31) becomes

$$\frac{dh}{dt} \sim \frac{S_{tri,m}}{S_{tri,m}+\varepsilon S_{strip}}\frac{\gamma\left[2\cos\theta+(\pi-\omega)\tan(\omega/2)\right]}{\left(1-\dfrac{\pi-\omega}{2}\tan(\omega/2)\right)\mu h}r_f - r_f^2\frac{S_{tri,m}}{S_{tri,m}+\varepsilon S_{strip}}\frac{\rho g\sin\alpha}{\mu}. \tag{6.32}$$

Here, the two terms in the right side of Eq. (6.32) can be integrated respectively, viz.

$$\frac{h^2}{2} \sim \frac{S_{tri,m}}{S_{tri,m}+\varepsilon S_{strip}}\frac{\gamma\left[2\cos\theta+(\pi-\omega)\tan(\omega/2)\right]}{\left(1-\dfrac{\pi-\omega}{2}\tan(\omega/2)\right)\mu h}r_f t, \tag{6.33}$$

and

$$h \sim -r_f^2 \frac{S_{tri,m}}{S_{tri,m} + \varepsilon S_{strip}} \frac{\rho g \sin\alpha}{\mu} t. \tag{6.34}$$

The integrated form of Eq. (6.32) can be obtained by considering Eq. (6.33) and Eq. (6.34),

$$h \sim \sqrt{\frac{2S_{tri,m}}{S_{tri,m} + \varepsilon S_{strip}} \frac{\gamma \left[2\cos\theta + (\pi - \omega)\tan(\omega/2)\right]}{\left(1 - \frac{\pi - \omega}{2}\tan(\omega/2)\right)\mu h} r_f t}$$
$$- r_f^2 \frac{2S_{tri,m}}{\rho \left(S_{tri,m} + \varepsilon S_{strip}\right)} \frac{\rho g \sin\alpha}{\mu} t. \tag{6.35}$$

When $\varepsilon S_{strip} \ll S_{tri,m}$ or within a short time, $\frac{2S_{tri,m}}{S_{tri,m} + \varepsilon S_{strip}}$ is close to a constant and Eq. (6.35) can be expressed as

$$h \sim \sqrt{A_3 r_f t - B_3 r_f^2 t}, \tag{6.36}$$

with $A_3 = \frac{2S_{tri,m}}{S_{tri,m} + \varepsilon S_{strip}} \frac{\gamma[2\cos\theta + (\pi-\omega)\tan(\omega/2)]}{\left(1 - \frac{\pi-\omega}{2}\tan(\omega/2)\right)\mu h}$ and $B_3 = \frac{2S_{tri,m}}{\rho\left(S_{tri,m} + \varepsilon S_{strip}\right)} \frac{\rho g \sin\alpha}{\mu}$.

Because the fluid front in the wedge stops rising from the boundary where the capillary pressure is equal to the gravitational pressure, we have $\frac{\partial h}{\partial r_f} = 0$ and then obtain the following equation from Eq. (6.36):

$$r_f \sim \frac{1}{t^{\frac{1}{3}}} \tag{6.37}$$

Substituting Eq. (6.37) into Eq. (6.36) leads to

$$t \sim h^3. \tag{6.38}$$

In the longer term, the maximum length or distance of the wetted area for drying is obtained at $\frac{dh}{dt} = 0$.

The microfibrous filter paper used is cellulose Whatman paper (Grade 1). The nanofibrous layer is made from nylon 6 pellets (CAS Number: 25038-54-4) purchased from Sigma-Aldrich. The solvent for solving nylon 6 pellets is formic acid (88%), which was purchased from Macron Fine Chemicals. Single-sided laminating sheets for bonding the fibrous layers were purchased from Scotch. The fluid liquid of wicking and drying is distilled water. For this distilled water in the lab with the standard atmosphere of 21 ± 2 °C and $62 \pm 2\%$ relative humidity, the surface tension γ is 7.235×10^{-2} N m^{-1}, the viscosity μ is 1.002×10^{-3} Pa m^{-1}, and the density ρ is 1.03 kg m^{-3}.

The nylon 6 pellets are dissolved in the formic acid at a concentration of 20 wt% by continuous stirring for 24 hours. Then the solution is added into a syringe with a metal needle and the flow rate is controlled by a digital flow pump. A high voltage is applied to the nylon 6 solution by connecting a copper wire to the metal tip. The voltage difference between the tip and the collector of the rotating metal drum used to electrospin the polymer solution is 20 kV. The drum is covered with metal-coated fabric from SEIREN CO. A distance of 15 cm from the spinning tip to the collector is set for collection of the electrospun nanofibers. The typical ejection rate of the solution is set at 0.5 mL h^{-1}. The nanofiber spinning is conducted in a fuming cupboard from 24 to 48 hours.

The porosity of each fibrous layered strip ε is measured based on the following correlation (Shou, Fan, et al. 2014): $\varepsilon = 1 - \rho_{porous} / \rho_{solid}$, where ρ_{solid} and ρ_{porous} are the densities of the solid materials and porous strips, respectively. As indicated by the product description of the filter paper, the mean pore radius is 11 µm. The mean pore radius of the fibrous layers is related to the porosity and the mean pore size using the Sampson's correlation for randomly distributed fiber networks (Sampson 2003), and is $R = -0.5\pi^{0.5} (1 + 0.5\pi / \ln\varepsilon) r_f$. Since the mean fiber radius r_f of the electrospun nanofibrous sheet is around 45 nm, the mean pore radius can be readily obtained at 554 nm. The thickness of the filter paper and the electropsun membrane is 0.17 mm and 0.10 m, respectively, measured by a MARATHON Digital Caliper.

The W-Channel is constructed in three steps (Figure 6.8). First, the laminating sheets are cut into uniform-width strips by a titanium rotary rotator trimmer and the filter papers are cut into three strip types and the three strips are labeled Component I, Component II, and Component III (Table 6.1). Second, Component III is stuck onto a sheet and Component II is packed onto the middle of Component III, both aligned with the strip of laminating sheet. Third, Component I is aligned on the top of both Component II and Component III and stuck by contact onto the laminating sheet. The layered strips of filter paper are bonded tightly to the laminating sheet, while Component II is fixed firmly between Components I and III. Thus, two triangular hollow wedged channels are generated between the three paper strips by virtue of the thickness of Component II. To fabricate a control sample, a single strip of Component I is fixed onto the laminating sheet. Analogously, the nanofiber-based W-Channels are constructed following the same procedure.

TABLE 6.1
Different W-Channel Types

Type of Device	Component I Material, width (mm)	Component II Material, width (mm)	Component III Material, width (mm)
A	Filter paper, 12.70	Filter paper, 3.18	Filter paper, 6.35
B	Filter paper, 12.70	None	None
C	Electrospun membrane, 12.70	Filter paper, 3.18	Electrospun membrane, 6.35
D	Electrospun membrane, 12.70	None	None
E	Filter paper, 63.50	Filter paper, 3.18 × 3	Filter paper, 6.35 × 3
F	Electrospun membrane, 38.10	Electrospun membrane, 3.18 × 2	Filter paper, 6.35 × 2

(Reprinted (adapted) with Permission from Ref. (Shou and Fan 2018). Copyright (2018) American Chemical Society)

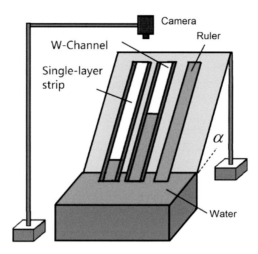

FIGURE 6.10 Schematic of experimental set-up for measuring dynamic capillary flow in W-Channels and single-layer strips.

A set-up is developed to examine the wicking dynamics for W-Channels placed at different incline angles (Figure 6.10). The water is gently injected into the liquid tank so that the bottoms of the W-Channels and the control samples are in the water to a depth of around 3 mm in the filled tank. Despite the complex process of the immediate water absorption by the samples, it occurs within a very short time (< 1 s), without delaying or accelerating the entire liquid transport process that commonly lasts more than 150 s. A volume-equivalent advancing line (i.e., the line normal to the flow direction that halves the area of the wetting parts within the rectangle that exactly contains the fronts of liquids with the minimum and maximum wicking distance) is used to determine the effective liquid fluid front. The wicking distance of the water is recorded at different times by taking camera snapshots periodically.

The detailed sample parameters of the single-layer control sample and the W-Channels are shown in Table 6.1. Figure 6.11a shows that the virtual appearance of the W-Channels is close to that of the control samples at the width of 1.27 cm. The scanning electron microscopy (SEM) images shown in Figure 6.11b provide clear demonstration of the microstructures of the samples, where both the nylon 6 electrospun nanofibers and the cellulose microfibers are randomly distributed in a nonwoven fashion. The diameters of the microfibers are between 5 μm and 30 μm, whereas the diameters of the nylon 6 nanofibers range from 50 nm to 100 nm. In Figure 6.11c, the geometrical structures of the triangular hollow wedge between the fibrous layers are also shown in the images of the cross-sectional W-Channel areas. To evaluate the wettability of the samples, tiny water droplets are dipped onto the surface of the electrospun membrane, the filter paper, and the laminating sheet and the contact angles are recorded by a contact angle goniometer (Rame-Hart 500). Figure 6.11d shows that both the filter paper and the electrospun membrane have contact angles smaller than 90° and the laminate film is hydrophobic with the contact angle at 124°. The spreading of water is much faster in the filter paper than in the electrospun membrane,

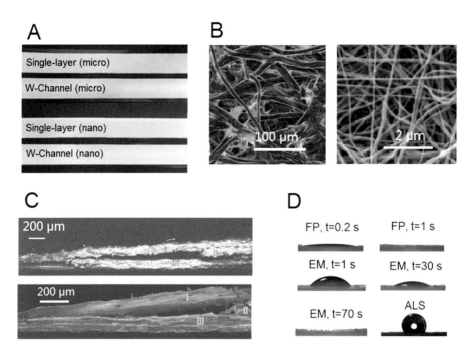

FIGURE 6.11 (a) Optical images of W-Channels and single-layer strips made of filter paper (micro) and electrospun membrane (nano). (b) SEM images of morphology of cellulose filter paper (left) and nylon 6 electrospun membrane (right). (c) SEM images of cross-sections of W-Channels made of filter paper (upper) and electrospun membrane (lower). (d) Contact angles of filter paper (FP), electrospun membrane (EM), and adhesive laminating sheet (ALS). (Reprinted with permission from D. Shou and J. Fan, "Design of nanofibrous and microfibrous channels for fast capillary flow." *Langmuir* 34 (2018): 1235–1241.)

which is ascribed to the fact that the former has much larger pores than the latter. The flow velocity is higher for the greater mean pore radius when the surface energies of the two samples are in the same order (Dourado et al. 1998).

Figure 6.12a shows the dynamic wicking processes in the filter paper W-Channel (Device A) and the single-layer strip (Device B) at various incline angles. It is demonstrated that it takes considerably less time for the liquid to wick in Device A than in Device B for the same flow distance, regardless of the incline angle of the samples. At 100 s, the wicking distance in Device A is 5 times greater than that in Device B when the two sample are layered horizontally. Even with the maximum gravitational effect, the wicking distance in Device A is more than twice than that in Device B at 100 s when they are vertically placed. When the gravitational force is ignored (when h is very small), the correlation between the wicking time and distance follows the Lucas-Washburn equation in both Device A and Device B, viz.,

$$t \sim h^2, \tag{6.39}$$

and Eq. (6.39) is validated by the log-log correlation summarized from the experimental data in Figure 6.12b. The channel of the hollow wedge is much larger

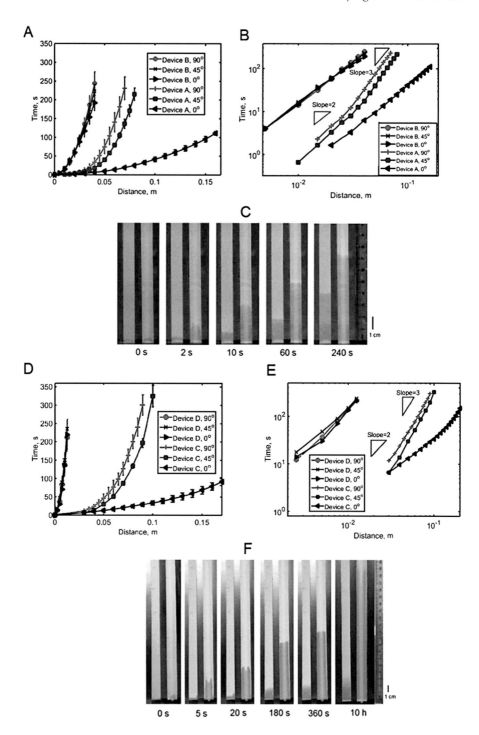

FIGURE 6.12 (*Continued*)

(*Continued*)

◀ **FIGURE 6.12**   Dynamics of wicking transport in W-Channels and single-layer strips. Comparison of capillary transport between W-Channels and single-layer strips made of microfibrous filter paper using (a) linear x-y axes and (b) log-log x-y axes under different incline angles. (c) Optical images of capillary rise in W-Channels and single-layer strips made of microfibrous filter paper at different times. Comparison of capillary transport between W-Channels and single-layer strips made of electrospun nanofibrous membrane using (d) linear x-y axes and (e) log-log x-y axes under different incline angles. (f) Optical images of capillary rise in W-Channels and single-layer strips made of electrospun nanofibrous membrane at different times. (Reprinted with permission from D. Shou and J. Fan, "Design of nanofibrous and microfibrous channels for fast capillary flow." *Langmuir* 34 (2018): 1235–1241.)

than the microspores in the porous fibrous layer, accounting for the faster wicking in Device A.

At different incline angles of $\alpha = 0°$, $\alpha = 45°$, and $\alpha = 90°$, the variation in wicking distance (when < 3.5 cm) is very small in Device B, indicating that the capillary force is much greater than the gravitational force considering the small pores in the filter paper and the evaporation loss is negligible. Thus all these flow processes are consistent with Eq. (6.39) at the early stage. It is also demonstrated that during the initial period, the time–distance correlations for Devices A are close to each other at different incline angles, with the slope of the log-log curve at 2 in Figure 6.12b. With the wicking process continuous, the advancing velocity decreases at the greater incline angle or with increasing gravitational effect. It is interesting to find that for a longer duration, the wicking time and distance in general follow cubic relationship, viz., $t \sim h^3$. The cubic relationship is evident by fitting the log-log plots in Figure 6.12b, a finding that agrees exactly with the analytical prediction in Eq. (6.38). During this time period, the effects of gravity and drying become more significant as the wicking process progresses. It is noted that the advance of the liquid between the larger region in the hollow wedged channel stops because of the balance between the capillary pressure and the gravitational pressure. However, the motion of the liquid in the narrower area continues, due to the higher capillary pressure, while the rising equilibrium meniscus approaches the corner edge of the wedge. This period is designated as Stage 2 and the previous period is designated as Stage 1, as characterized by Eq. (6.38) and Eq. (6.39), respectively. Furthermore, the flow resistance of the hollow wedge channel partially filled with the water remains small because of the larger pores compared to the microspores in Device B. Therefore, the fast wicking transport remains in Device A by adaptively benefiting from the increase in capillary pressure while the viscous resistance is still relatively low. Faster wicking against gravity is more clearly demonstrated in Device A than in Device B along with time, as seen in Figure 6.12c.

We also compared the capillary flows at different incline angles along the W-Channel made of nanofibrous membranes (Device C) with those along the single-layer strip (Device D). As seen in Figure 6.12d and Figure 6.12e, the general trend of capillary transport is similar to that in the W-Channel made of filter papers, but the flow enhancement of the W-Channel compared with the control sample is much greater. This phenomenon is attributed to the fact that the flow resistance of nano-fibrous membranes is extremely high, leading to very slow flow, but it rarely influences the fast capillary flow within the hollow wedges that governs the overall fluid advance. Therefore, the W-Channel has great potential for fluidic applications based on nanoscale materials such as thin-film evaporative cooling systems and nanofiber-based diagnostics.

In Figure 6.12f, the optical images show the significant differences in the flow distance between Device C and Device D composed of nanofibers at different times. After a longer duration, the effect of drying becomes important, further slowing the capillary rise. The final height of the liquid front is reached in the two devices as a result of the balance of capillary and gravitational pressures, taking evaporation loss into consideration. It is noted that the final height of the fluid front in Device C is approximately 19 cm, whereas that in Device D is only around 3 cm at $t = 10$ hour (Figure 6.12f). The significantly increased height in Device C is caused by the higher flow rate and the greater amount of water remaining in the hollow wedge as a source for evaporation, in comparison to the single-layer case. When drying can be ignored, the final height is highly dependent on the (minimum) size of the porous systems. In this condition, we believe that the final height will be equal in the W-Channel and the single-layer strip.

6.5 SUMMARY REMARKS

This chapter summarizes theoretical models of heat and mass transfer in drying and particularly the wicking and drying process in the fibrous porous systems. To overcome the bottleneck of limited wicking under the effects of drying and gravity, an optimized fibrous medium and a W-Channel composed of nano- or microfibrous layers are developed with significantly improved ability for liquid delivery. In the W-Channels the flow process follows $t \sim h^2$ or the Lucas-Washburn equation in the early stage, but in the long run, the capillary rise follows $t \sim h^3$. The enhancement of the wicking process is greater in the nano-fiber-based W-Channels. Based on the derived models, the optimized pore sizes are also found to maximize drying by properly controlling the wicking process in the proposed fibrous porous media. Therefore, the model-driven proposed fibrous structure has huge potential for a variety of evaporative drying-based applications including microfluidic diagnostics, large-scale thermal management, solar steam generation, and functional textiles.

NOMENCLATURE

a	Thickness	$C_v(x, t)$	Effective volumetric heat capacity of the fibrous batting
A	Constant		
A_{SL}	Solid–liquid interfacial area	$C_{vv}(x, t)$	Effective volumetric heat capacity of the water vapor
A_{SV}	Solid–vapor interfacial area		
A_{VL}	Vapor–liquid interfacial area		

D_a	Diffusion coefficient of water vapor in the air	S_{tri}	Area of triangular area
		t	Flow time
D_1	Diffusion coefficient of liquid water	T	Temperature
		T_s	Temperature at the interface between vapor and liquid
E	Evaporation or condensation coefficient		
		T_v	Water vapor temperature in the air
F_L	Thermal radiation incident traveling to the left		
		W	Water content of the fibrous layer
F_R	Thermal radiation incident traveling to the right		
		α	Incline angle
g	Gravitational acceleration	β	Absorption constant
h	Flow distance	γ	Surface tension
h_{max}	Maximum distance or height	γ_{SL}	Solid–liquid interfacial tension
$k(x, t)$	Effective thermal conductivity		
		γ_{SV}	Solid–vapor interfacial tension
k_e	Evaporation rate per area		
K_p	Hydraulic permeability	γ_{VL}	Vapor–liquid interfacial tension
K_t	Permeability of the channel		
l	Flow distance of wetted area	ε	Porosity of the fibrous medium
M	Molecular weight of water		
p_c	Capillary pressure	ε_s	Porosity of the porous strip
p_g	Gravitational pressure	θ	Contact angle
P_{sat}	Saturated water vapor pressure	$\lambda(x, t)$	Latent heat of water
		μ	Viscosity of the liquid
P_v	Water vapor pressure in the air	ρ	Fluid density
Δp	Viscous pressure	σ	Boltzmann constant
Q	Effective flow rate	τ	Tortuosity factor
r_f	Radius of the curvature of the fluid front	$\Gamma_{ce}(x, t)$	Rate of condensation or evaporation
r_{in}	Inscribed radius of the wedge channel	$\Gamma(x, t)$	Rate of evaporation or condensation
R	Tube radius		
R_u	Universal gas constant	μ	Viscosity of the liquid
S_{strip}	Total cross-sectional area		

REFERENCES

Adera, S., D. Antao, R. Raj, and E. N. Wang. 2016. "Design of micropillar wicks for thin-film evaporation." *International Journal of Heat and Mass Transfer* 101:280–294. doi: 10.1016/j.ijheatmasstransfer.2016.04.107.

Alavudeen, A., M. Thiruchitrambalam, N. Venkateshwaran, and A. Athijayamani. 2011. "Review of natural fiber reinforced woven composite." *Reviews on Advanced Materials Science* 27 (2):146–150.

Bergles, A. E., J. H. Lienhard, G. E. Kendall, and P. Griffith. 2003. "Boiling and evaporation in small diameter channels." *Heat Transfer Engineering* 24 (1):18–40. doi: 10.1080/01457630390116086.

Butt, H.-J., K. Graf, and M. Kappl. 2006. *Physics and Chemistry of Interfaces.* Weinheim: Wiley-VCH.

Cai, J. C., X. Y. Hu, D. C. Standnes, and L. J. You. 2012. "An analytical model for spontaneous imbibition in fractal porous media including gravity." *Colloids and Surfaces a-Physicochemical and Engineering Aspects* 414:228–233. doi: 10.1016/j.colsurfa.2012.08.047.

Cai, J. C., E. Perfect, C. L. Cheng, and X. Y. Hu. 2014. "Generalized modeling of spontaneous imbibition based on Hagen-Poiseuille flow in tortuous capillaries with variably shaped apertures." *Langmuir* 30 (18):5142–5151. doi: 10.1021/la5007204.

Cai, Q. J., and A. Bhunia. 2012. "High heat flux phase change on porous carbon nanotube structures." *International Journal of Heat and Mass Transfer* 55 (21–22):5544–5551. doi: 10.1016/j.ijheatmasstransfer.2012.05.027.

Camplisson, Conor K., Kevin M. Schilling, William L. Pedrotti, Howard A. Stone, and Andres W. Martinez. 2015. "Two-ply channels for faster wicking in paper-based microfluidic devices." *Lab on a Chip* 15 (23):4461–4466. doi: 10.1039/c5lc01115a.

Chen, Huawei, Pengfei Zhang, Liwen Zhang, Hongliang L. Iu, Ying Jiang, Deyuan Zhang, Zhiwu Han, and Lei Jiang. 2016. "Continuous directional water transport on the peristome surface of Nepenthes alata." *Nature* 532 (7597):85-+. doi: 10.1038/nature17189.

Chen, W., S. Liu, and J. Lin. 2015. "Analysis on the passive evaporative cooling wall constructed of porous ceramic pipes with water sucking ability." *Energy and Buildings* 86:541–549. doi: 10.1016/j.enbuild.2014.10.055.

Chen, Zuo-Rong, Lin Ye, and Teresa Kruckenberg. 2006. "A micromechanical compaction model for woven fabric preforms. Part I: Single layer." *Composites Science and Technology* 66 (16):3254–3262. doi: 10.1016/j.compscitech.2005.07.028.

Dourado, F., F. M. Gama, E. Chibowski, and M. Mota. 1998. "Characterization of cellulose surface free energy." *Journal of Adhesion Science and Technology* 12 (10):1081–1090. doi: 10.1163/156856198x00740.

Fan, J., X. Cheng, X. Wen, and W. Sun. 2004. "An improved model of heat and moisture transfer with phase change and mobile condensates in fibrous insulation and comparison with experimental results." *International Journal of Heat and Mass Transfer* 47 (10–11):2343–2352. doi: 10.1016/j.ijheatmasstransfer.2003.10.033.

Fan, J., Z. Luo, and Y. Li. 2000. "Heat and moisture transfer with sorption and condensation in porous clothing assemblies and numerical simulation." *International Journal of Heat and Mass Transfer* 43 (16):2989–3000. doi: 10.1016/s0017-9310(99)00235-5.

Fan, J., and X. Wen. 2002. "Modeling heat and moisture transfer through fibrous insulation with phase change and mobile condensates." *International Journal of Heat and Mass Transfer* 45 (19):4045–4055. doi: 10.1016/s0017-9310(02)00114-x.

Ghasemi, H., G. Ni, A. M. Marconnet, J. Loomis, N. Yerci, N. Miljkovic, and G. Chen. 2014. "Solar steam generation by heat localization." *Nature Communications* 5:4449. doi: 10.1038/ncomms5449.

Giokas, Dimosthenis L., George Z. Tsogas, and Athanasios G. Vlessidis. 2014. "Programming fluid transport in paper-based microfluidic devices using razor-crafted open channels." *Analytical Chemistry* 86 (13):6202–6207. doi: 10.1021/ac501273v.

Jahanshahi-Anbuhi, Sana, Puneet Chavan, Clemence Sicard, Vincent Leung, S. M. Zakir Hossain, Robert Pelton, John D. Brennan, and Carlos D. M. Filipe. 2012. "Creating fast flow channels in paper fluidic devices to control timing of sequential reactions." *Lab on a Chip* 12 (23):5079–5085. doi: 10.1039/c2lc41005b.

Jia, P., M. Z. Dong, L. M. Dai, and J. Yao. 2008. "Slow viscous flow through arbitrary triangular tubes and its application in modelling porous media flows." *Transport in Porous Media* 74 (2):153–167. doi: 10.1007/s11242-007-9187-3.

Li, X. Q., W. C. Xu, M. Y. Tang, L. Zhou, B. Zhu, S. N. Zhu, and J. Zhu. 2016. "Graphene oxide-based efficient and scalable solar desalination under one sun with a confined 2D water path." *Proceedings of the National Academy of Sciences of the United States of America* 113 (49):13953–13958. doi: 10.1073/pnas.1613031113.

Liu, M. C., J. Wu, Y. X. Gan, D. A. H. Hanaor, and C. Q. Chen. 2016. "Evaporation limited radial capillary penetration in porous media." *Langmuir* 32 (38):9899–9904. doi: 10.1021/acs.langmuir.6b02404.

Mariatti, M., M. Nasir, and H. Ismail. 2000. "Effect of sample cutting direction on mechanical properties of woven thermoplastic prepreg." *Polymer Testing* 19 (6):617–624. doi: 10.1016/s0142-9418(99)00032-x.

Motakef, S., and M. A. Elmasri. 1986. "Simultaneous heat and mass-transfer with phase-change in a porous slab." *International Journal of Heat and Mass Transfer* 29 (10):1503–1512. doi: 10.1016/0017-9310(86)90065-7.

Ni, G., G. Li, S. V. Boriskina, H. X. Li, W. L. Yang, T. J. Zhang, and G. Chen. 2016. "Steam generation under one sun enabled by a floating structure with thermal concentration." *Nature Energy* 1:16126. doi: 10.1038/nenergy.2016.126.

Ogniewicz, Y., and C. L. Tien. 1981. "Analysis of condensation in porous insulation." *International Journal of Heat and Mass Transfer* 24 (3):421–429. doi: 10.1016/0017-9310(81)90049-1.

Ponomarenko, Alexandre, David Quere, and Christophe Clanet. 2011. "A universal law for capillary rise in corners." *Journal of Fluid Mechanics* 666:146–154. doi: 10.1017/s0022112010005276.

Reyssat, M., L. Courbin, E. Reyssat, and H. A. Stone. 2008. "Imbibition in geometries with axial variations." *Journal of Fluid Mechanics* 615:335–344. doi: 10.1017/s0022112008003996.

Sampson, W. W. 2003. "A multiplanar model for the pore radius distribution in isotropic near-planar stochastic fibre networks." *Journal of Materials Science* 38 (8):1617–1622. doi: 10.1023/a:1023298820390.

Shou, D. H., and J. T. Fan. 2015a. "The fastest capillary penetration of power-law fluids." *Chemical Engineering Science* 137:583–589. doi: 10.1016/j.ces.2015.07.009.

Shou, D. H., and J. T. Fan. 2018. "Design of nanofibrous and microfibrous channels for fast capillary flow." *Langmuir* 34 (4):1235–1241. doi: 10.1021/acs.langmuir.7b01797.

Shou, D. H., and J. T. Fan. 2015b. "Structural optimization of porous media for fast and controlled capillary flows." *Physical Review E* 91 (5):053021. doi: 10.1103/PhysRevE.91.053021.

Shou, Dahua, Jintu Fan, Maofei Mei, and Feng Ding. 2014. "An analytical model for gas diffusion though nanoscale and microscale fibrous media." *Microfluidics and Nanofluidics* 16 (1–2):381–389. doi: 10.1007/s10404-013-1215-8.

Shou, Dahua, Lin Ye, and Jintu Fan. 2014a. "The fastest capillary flow under gravity." *Applied Physics Letters* 104 (23):231602. doi: 10.1063/1.4882057.

Shou, Dahua, Lin Ye, and Jintu Fan. 2014b. "Treelike networks accelerating capillary flow." *Physical Review E* 89 (5):053007. doi: 10.1103/PhysRevE.89.053007.

Shou, Dahua, Lin Ye, Jintu Fan, and Kunkun Fu. 2014. "Optimal design of porous structures for the fastest liquid absorption." *Langmuir* 30 (1):149–155. doi: 10.1021/la4034063.

Tomadakis, M. M., and T. J. Robertson. 2005. "Viscous permeability of random fiber structures: Comparison of electrical and diffusional estimates with experimental and analytical results." *Journal of Composite Materials* 39 (2):163–188. doi: 10.1177/0021998305046438.

Washburn, E. W. 1921. "The dynamics of capillary flow." *Physicak Review* 17 (3):273–283. doi: 10.1103/PhysRev.17.273.

Xiao, Junfeng, Howard A. Stone, and Daniel Attinger. 2012. "Source-like solution for radial imbibition into a homogeneous semi-infinite porous medium." *Langmuir* 28 (9):4208–4212. doi: 10.1021/la204474f.

Xiao, R., S. C. Maroo, and E. N. Wang. 2013. "Negative pressures in nanoporous membranes for thin film evaporation." *Applied Physics Letters* 102 (12):123103 doi: 10.1063/1.4798243.

Xu, N., X. Z. Hu, W. C. Xu, X. Q. Li, L. Zhou, S. N. Zhu, and J. Zhu. 2017. "Mushrooms as efficient solar steam-generation devices." *Advanced Materials* 29 (28):201606762. doi: 10.1002/adma.201606762.

Xu, P., X. L. Ma, X. D. Zhao, and K. S. Fancey. 2016. "Experimental investigation on performance of fabrics for indirect evaporative cooling applications." *Building and Environment* 110:104–114. doi: 10.1016/j.buildenv.2016.10.003.

7 Capillary Valve Effect on Drying of Porous Media

R. Wu
Shanghai Jiao Tong University, Shanghai, China

C.Y. Zhao
Shanghai Jiao Tong University, Shanghai, China

E. Tsotsas
Otto von Guericke University, Magdeburg,
Sachsen-Anhalt, Germany

A. Kharaghani
Otto von Guericke University, Magdeburg,
Sachsen-Anhalt, Germany

CONTENTS

7.1 INTRODUCTION

Two-phase transport in porous media is of great interest to many industrial fields, such as CO_2 sequestration, recovery of volatile hydrocarbons from underground oil reservoirs, remediation of contaminated soils by vapor extraction, and water management of proton exchange membrane fuel cells. Nevertheless, it is a challenge to fully understand the two-phase transport in a porous material since it is affected not only by interactions between gravitational, capillary, and viscous forces but also by the structure of the pore space.

Porous materials contain pores of various sizes such that small pores may be connected to large pores with a sudden geometrical expansion at their interfaces. Here, the small pore is called the pore throat, and the large pore is called the pore body. The sudden geometrical expansion can increase the resistance to the advancement of the invading fluid and has already been employed as a capillary valve to control the fluid flow in microfluidic devices (Duffy et al., 1999; Cho et al., 2007; Chen et al., 2008; Moore et al., 2011). When the invading fluid reaches a sudden geometrical expansion between a pore throat and a pore body, the contact angle is suddenly increased, thereby hindering the movement of the three-phase contact line. The three-phase contact line will advance again if the pressure of the invading fluid increases to a critical value. From this point of view, the sudden geometrical expansion between a pore throat and a pore body serves as a valve that controls meniscus movement. This is called the capillary valve effect (CVE). Such CVE will influence the menisci movement and hence the two-phase transport in porous media.

To reveal the influence of the CVE on two-phase transport in porous media, a pore-scale investigation is necessary. It is nontrivial to capture experimentally the pore-scale events during two-phase transport in real porous materials since they are not only opaque but also have complex microstructures. An alternative is to use the pore network (PN) modeling approach. This approach has been widely used as an effective tool to understand the two-phase transport in porous media (Blunt, 2001; Joekar-Niasar and Hassanizadeh, 2012). In this method, the void space of a porous medium is conceptualized as a pore network composed of regular pores of various sizes. The two-phase transport in a PN is depicted by the prescribed rules. For instance, to simulate the capillary force–dominated two-phase flow in porous media using the PN model, the invasion percolation algorithm proposed by Wilkiinson and Willemsen (1983) has been widely used (Blunt et al., 1992; Knackstedt et al., 1998, 2001; Mani and Mohanty, 1999; Lopez et al., 2003; Araujo et al., 2005; Ceballos and Prat, 2013).

In the invasion percolation algorithm for two-phase flow in a PN, only one pore is invaded at each step. This invaded pore is the largest available one in the drainage process, whereas in the imbibition process, it is the smallest available one. Bazylak et al. (2008) compared the PN simulations against the experiments for slow drainage in PNs of various structures. Differences between the experimental and simulation results were observed and attributed to the uncertainty in the fabrication of the PN. Although fabrication uncertainty is an important reason, it will be shown later that neglecting the CVE in the PN model can be another reason. This also indicates that the CVE must be taken into account in the PN models.

In this chapter, we will show the influence of the CVE on the two-phase transport, especially drying, in porous media based on the PN modeling approach. Various PN models have been developed to investigate effects on drying of porous media of the pore structures (Metzger et al., 2007; Pillai et al., 2009), viscous forces (Metzger et al., 2008; Taleghani and Dadvar, 2014), thermal gradient (Plourde and Prat, 2003; Surasani et al., 2008), mechanical damage (Kharaghani et al., 2011 and 2012), and liquid films (Yiotis et al., 2004; Vorhauer et al., 2015). However, in these PN studies, the CVE is not considered. If the CVE is considered in the PN model, the modeling results have better agreement with the experimental data (Wu et al., 2016a). Hence, the CVE must be considered so as to understand in detail drying of porous media.

In what follows, the CVE is explained in detail, and two types of invasion into pore body are introduced. In Section 7.3, the PN model with the CVE for capillary force–dominated two-phase flow in porous media is introduced. Based on this two-phase flow PN model, the PN model with the CVE is developed for slow drying of porous media in Section 7.4. Based on the PN model for slow drying, the PN model with the CVE for convective drying in porous media is introduced in Section 7.5. The summarization is presented in Section 7.6.

7.2 CAPILLARY VALVE EFFECT

The void space in a porous material is composed of pores of various sizes. At the interface between a small pore (pore throat) and a large pore (pore body), a sudden geometrical expansion structure may exist. Such sudden geometrical expansion will influence movement of menisci during two-phase flow in a porous material. To understand clearly the effect of sudden geometrical expansion on movement of menisci, we first introduce the quasi-static two-phase displacement in a pore with a rectangular cross section. The invading and displaced fluids are separated by a meniscus, across which a pressure difference is established:

$$P_I - P_D = \sigma \left(\frac{1}{r_w} + \frac{1}{r_h} \right) \tag{7.1}$$

where P_I and P_D are the pressure of the invading and displaced fluid, respectively; σ is the surface tension; and r_w and r_h are the radii of curvatures of the menisci in the width and height directions, respectively. These two curvature radii have direction as well as magnitude. The radius of a curvature is taken as positive if the curvature center is at the side of the invading fluid, otherwise negative. For the meniscus shown in Figure 7.1, both the curvature radii are positive, and $r_h = h/2\cos\theta$, and $r_w = w/2\cos\theta$, where θ is the contact angle, and h and w are the channel height and

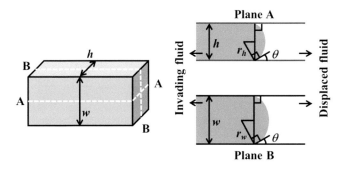

FIGURE 7.1 The shape of a meniscus in a pore.

width, respectively. The contact angle is defined as the angle between the interface of the invading fluid and the displaced fluid and the interface of the displaced fluid and the solid. In the drying case, liquid in porous media is gradually replaced by the gas phase, similar to gas invasion; for such circumstance, gas is the invading fluid, and liquid is the displaced fluid.

For two-phase flow in a pore, the three-phase contact line cannot move when the contact angle is larger than the advancing contact angle θ_a. The value of the contact angle is relevant to the pressure difference across the meniscus between the invading fluid and the displaced fluid, since the radius of the meniscus curvature depends on the contact angle, Eq. (7.1). The lower the contact angle is, the higher the pressure difference across the meniscus. To advance a meniscus in a pore, the pressure difference across the meniscus must be larger than a critical value so that the contact angle is equal to or smaller than the advancing contact angle. This critical value is called the threshold pressure. Obviously, the threshold pressure for a pore of height h and width w is:

$$P_{th} = 2cos\theta_a \sigma \left(\frac{1}{w} + \frac{1}{h} \right) \tag{7.2}$$

Eq. (7.2) can be used to determine the threshold pressure for invasion into a pore throat from a pore body. But, for invasion from a pore throat into a pore body with a sudden geometrical expansion, Eq. (7.2) is not applicable. Figure 7.2 shows the process of the invading fluid entering a pore body from a pore throat. The pore throat and the pore body have the same height but different widths. The height direction is perpendicular to the plane shown in Figure 7.2. When the three-phase contact line moves from the pore throat to the interface between the pore throat and the pore body, the contact angle jumps from θ_a to $\theta_a + 90°$ due to the sudden

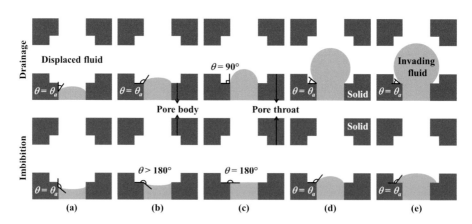

FIGURE 7.2   Schematic of bursting invasion into a pore body from a pore throat: (a) advancement of the three-phase contact line in the pore throat; (b–d) evolution of the meniscus when the three-phase contact line is pinned at the interface between the pore throat and the pore body; (e) advancement of the three-phase contact line along the wall of the pore body.

geometrical expansion, Figures 7.2a and 7.2b. To this end, the three-phase contact line is pinned, and the radius of the meniscus curvature along the width direction is $r_w = w_t/2\sin\theta$, where w_t is the width of the pore throat. As the invading fluid flows, the meniscus grows, and the contact angle reduces from $\theta = \theta_a + 90°$ to θ_a, Figures 7.2b–7.2d. During this period, r_w is varied, but the radius of the meniscus curvature along the height direction remains unchanged $r_h = h/2\cos\theta_a$ since the pore throat and the pore body have the same height h. To this end, as the contact angle varies from $\theta = \theta_a + 90°$ to θ_a, the pressure difference across the meniscus, determined by Eq. (7.1), is maximum at $\theta = \max(90°, \theta_a)$, i.e., $90°$ in drainage of a non-wetting fluid displacing a wetting fluid, and θ_a ($>90°$) in imbibition of a wetting fluid displacing a non-wetting fluid. As the contact angle reduces to θ_a, the three-phase contact line begins to move along the wall of the pore body, Figures 7.2d and 7.2e. The radius of meniscus curvature along the width direction then increases, resulting in a lower pressure difference across the meniscus, Eq. (7.1). Hence, the threshold pressure for invasion into a pore body from a pore throat with the same height h is:

$$P_{th} = 2\sigma \left(\frac{sin\left[max\left(90°, \theta_a\right)\right]}{w_t} + \frac{cos\theta_a}{h} \right) \tag{7.3}$$

where w_t is the width of the pore throat.

Because of the sudden geometrical expansion at the interface between the pore throat and the pore body, the threshold pressure to invade a pore throat is smaller than that to burst from it into a pore body, Eqs. (7.2) and (7.3). It seems that such geometrical expansion serves as a valve to control the movement of the invading fluid. This is called the capillary valve effect (CVE). During two-phase flow in a porous material, two or more pore throats adjacent to a pore body can be invaded before invasion into this pore body.

Figure 7.3 shows the process of invading fluid entering a pore body from two adjacent pore throats. Pore throats A and B are invaded, and $w_A > w_B$. All the pores have the same height h. Menisci A and B have the same curvature radii since the pressure differences across them are identical when two-phase invasion is dominated by capillary forces. This also implies $\theta_A > \theta_B$. The menisci are

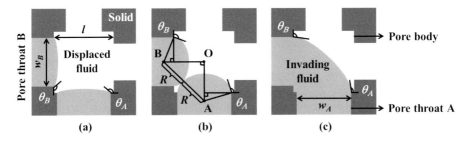

(a)   **(b)**   **(c)**

FIGURE 7.3 Schematic of merging invasion into a pore body: (a) before merging of two menisci; (b) at the moment of merging; (c) after merging of two menisci.

merged at $\theta_A > \max(90°, \theta_a)$. The curvature radii are $r_w = w_A/2\sin\theta_A$ before merging, Figures 7.3a and 7.3b. After merging, a new meniscus forms, leading to a larger r_w, Figure 7.3c. The pressure difference across the menisci is largest at the moment of merging, Eq. (7.1). This type of invasion into a pore body is called merging invasion. The threshold pressure is:

$$P_{th} = \sigma\left(\frac{1}{R} + \frac{2cos\theta_a}{h}\right) \tag{7.4}$$

where R is the radius of curvature along the width direction at the moment of menisci merging. As elucidated in Figure 7.3b, the value of R can be determined by the relationship $AB^2 = OA^2 + OB^2$:

$$(2R)^2 = \left[\frac{l}{2} + \sqrt{R^2 - \left(\frac{w_A}{2}\right)^2}\right]^2 + \left[\frac{l}{2} + \sqrt{R^2 - \left(\frac{w_B}{2}\right)^2}\right]^2 \tag{7.5}$$

At the moment of menisci merging, θ_A is larger than the value of $\max(90°, \theta_a)$, which implies:

$$R < \frac{w_A}{2sin\left[\max\left(90°, \theta_a\right)\right]} \tag{7.6}$$

The menisci cannot merge at $\theta_A \geq 145°$ since the pore bodies are square and larger than their adjacent pore throats. This means:

$$R > \frac{w_A}{\sqrt{2}} \tag{7.7}$$

The value of R can be determined from Eqs. (7.5)–(7.7). If no solution is found, the merging invasion shown in Figure 7.3 will not occur; and the pore body will be invaded by bursting invasion from pore throat A.

For the merging invasion into a pore body shown in Figure 7.3, two menisci attached to this pore body are merged. Nevertheless, when two menisci are merged during invasion into a pore body, the invasion can also be bursting invasion. To understand bursting and merging invasion clearly, we compare in Figure 7.4 four cases of invasion into a pore body.

The pores have the same height but different widths. Both pore throats A and B are occupied by the invading fluid, and pore throat A is larger than pore throat B. In Figures 7.4a and 7.4c, the pores are hydrophobic with an advancing contact angle of θ_{ho}, and the invading fluid is the wetting phase; while in Figures 7.4b and 7.4d, the pores are hydrophilic with an advancing contact angle of θ_{hi}, and the invading fluid is a non-wetting phase. Although menisci are merged in all cases shown in Figure 7.4, invasion shown in Figures 7.4a and 7.4b is bursting invasion, whereas invasion shown in Figures 7.4c and 7.4d is merging invasion.

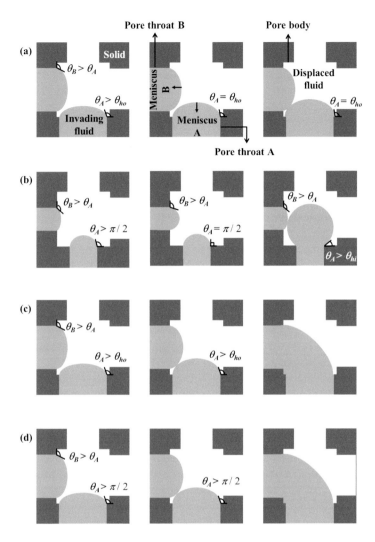

FIGURE 7.4 Invasion into a pore body with menisci merging; (a) bursting invasion into a hydrophobic pore body; (b) bursting invasion into a hydrophilic pore body; (c) merging invasion into a hydrophobic pore body; (d) merging invasion into a hydrophilic pore body. The invading fluid is gas, and the displaced fluid is liquid.

For invasion into the hydrophobic pore body shown in Figure 7.4a, contact angles of both menisci first increase. The contact angel of meniscus A is smaller than that of meniscus B since pore throat A is larger. When the contact angle of meniscus A reaches the advancing contact angle of the pore body (see the middle image of Figure 7.4a), the three-phase contact line of meniscus A starts moving along the wall of the pore body, and then the two menisci are merged (see the right image of Figure 7.4a). It should be noted that in the middle image of Figure 7.4a, the pressure difference across the menisci is equal to the threshold pressure for bursting into the pore body from pore throat A, Eq. (7.3). For this reason, the invasion shown in Figure 7.4a is still considered as bursting invasion. If menisci A and B are merged before the contact angle of meniscus A reaches the advancing contact angle of pore body (as that shown in Figure 7.4c), then invasion is merging invasion. In this merging invasion, the threshold pressure across menisci is always smaller than the threshold pressure for bursting invasion into the pore body from pore throat A. The threshold pressure for bursting invasion into the pore body from pore throat A is smaller than that from pore throat B since pore throat A is larger, Eq. (7.3).

For invasion into the hydrophilic pore body shown in Figure 7.4b, the two menisci are merged before the three-phase contact line of any meniscus move along the wall of the pore body. But, at the merging moment, the contact angle of meniscus A is acute, see the right image of Figure 7.4b. For this invasion, the pressure difference across the menisci is equal to the threshold pressure for bursting invasion into the pore body from the pore throat A when the contact angle of meniscus A reduces to the right angle, see the middle image of Figure 7.4b. In this regard, the invasion shown in Figure 7.4b is still bursting invasion. If the menisci are merged at the moment when the contact angle of meniscus A is obtuse, as shown in Figure 7.4d, then the invasion is merging invasion.

For the merging invasion shown in Figures 7.3 and 7.4, the two menisci to be merged neighbor each other. If they were opposite, merging invasion would not occur since the pore body is square (in the plane perpendicular to the height direction) and larger than the pore throats. Three or more menisci attached to a pore body cannot be merged simultaneously since pore throats have various sizes. As a result, only merging invasion into a pore body by two connected menisci neighboring to each other is considered.

The following procedure is used to check whether a pore body is invaded by bursting or merging invasion. First, the invaded pore throats adjacent to the pore body are determined, and the meniscus attached to the largest invaded pore throat is called meniscus A. The menisci neighboring meniscus A are then scanned. If there is no such meniscus, the pore body will be invaded by bursting invasion. If there are menisci neighboring meniscus A, the one attached to the pore throat of the largest size is identified and called meniscus B. The possibility of merging invasion by menisci A and B is then checked by Eqs. (7.5)–(7.7). Merging invasion will happen if a solution is found; otherwise bursting invasion of meniscus A will occur.

In the above analysis, the threshold pressure for bursting invasion into a pore body is gained based on the evolution of the meniscus observed experimentally. To derive the threshold pressure for bursting invasion theoretically, we consider invasion from a cylindrical pore throat into a cubic pore body, as illustrated in Figure 7.5. Initially, the three-phase contact line is located at the interface between the pore throat and the pore body. As the invading fluid is injected, the shape of the meniscus varies, resulting in a reduced contact angle. The dashed lines shown in Figure 7.5 elucidate the variation of the meniscus shape during invasion. The three-phase contact line is pinned until the contact angle reduces to the value of the advancing contact angle.

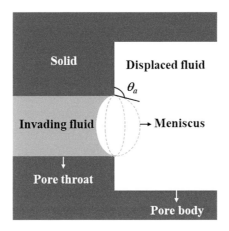

FIGURE 7.5   Schematic of a bursting invasion into a cubic pore body from a cylindrical pore throat. The dashed lines elucidate the evolution of the meniscus shape during invasion.

As the volume of the invading fluid increases by an infinitesimal value ΔV, the work done to the meniscus with the pinned three-phase contact line is:

$$W_m = (P_I - P_D)\Delta V \qquad (7.8)$$

where P_I and P_D are the pressures in the invading and displaced fluids, respectively. The work done to the meniscus is equal to the change of its interfacial energy, which is expressed as:

$$\Delta U_T = \sigma \Delta A_m \qquad (7.9)$$

where A_m is the interfacial area of the meniscus. Combing Eqs. (7.8) and (7.9), we can get:

$$P_I - P_D = \sigma \frac{dV}{dA_m} \qquad (7.10)$$

Since the shape of the meniscus is spherical, the volume of the invading fluid in the system is:

$$V = V_t + \frac{\pi r^3}{3} \frac{(1+cos\theta)^2 (2-cos\theta)}{sin^3\theta} \qquad (7.11)$$

where θ is the contact angle; V_t and r are the volume and radius of the pore throat, respectively. The interfacial area of the meniscus is:

$$A_m = \frac{2\pi r^2}{1-cos\theta} \qquad (7.12)$$

To this end, the pressure difference across the meniscus is determined as:

$$P_I - P_D = \frac{2\sigma sin\theta}{r}$$

(7.13)

Eq. (7.13) indicates that during the invasion shown in Figure 7.5, the pressure difference across the meniscus peaks at $\theta = 90°$ if the advancing contact angle of the pore body is $\theta_a \le 90°$ but at $\theta = \theta_a$ if $\theta_a > 90°$. Hence, the threshold pressure for bursting into a pore body from its neighboring pore throat of radius r_t is:

$$P_{th} = \frac{2\sigma \sin\left[\max\left(90°, \theta_a\right)\right]}{r_t}$$

(7.14)

Similarly, the threshold pressure of a cylindrical pore throat of radius r_t can also be theoretically derived as $P_{th} = \frac{2\sigma cos\theta_a}{r_t}$. The threshold pressure for merging invasion into a cubic pore body from two neighboring cylindrical pore throats (denoted as A and B) is therefore $P_{th} = \frac{2\sigma}{R}$, where R is determined by the equation of $(2R)^2 = \left[r_b + \sqrt{R^2 - r_{t,A}^2}\right]^2 + \left[r_b + \sqrt{R^2 - r_{t,B}^2}\right]^2$. The value of R should be $R > \frac{\max(r_{t,A}, r_{t,B})}{\sin\left[\max(90°, \theta_a)\right]}$ and $R < \sqrt{2}\max\left(r_{t,A}, r_{t,B}\right)$. If no solution is gained for R, then a merging invasion is impossible. Here, r_b and r_t represent the radii of pore body and throat, respectively. The radius of a cubic pore body is defined as the radius of the largest sphere that can be inscribed in the pore body.

7.3 CAPILLARY VALVE EFFECT ON TWO-PHASE FLOW IN POROUS MEDIA

In the foregoing section, we introduced the CVE. In this section, the PN model with the CVE for two-phase flows in porous media (Wu et al., 2016a) is introduced. To validate the developed PN model, experiments are performed for gas-liquid two-phase flows in microfluidic PNs.

7.3.1 Two-phase Flow Experiment

The gas-liquid two-phase flow experiments are conducted with the microfluidic PNs supplied by CapitalBio Corporation (China). The transparent PNs are fabricated using PDMS and have a semi-two-dimensional structure. The pores have the same depth of $h = 0.1$ mm. In the plane perpendicular to the depth direction, the PNs consist of square pore bodies with the side length of $l = 1$ mm and of rectangular pore throats with a randomly distributed width w; the distance between the centers of two neighboring pore bodies is $a = 2$ mm.

Two types of PNs with different throat width distributions are used. In the PN of type A, the pore throat widths are uniformly distributed in the range 0.14–0.94 mm. The minimal difference between two throat widths is 0.02 mm so as to relieve the effects of the fabrication uncertainty (±0.01 mm). In the PN of type B, the pore throat

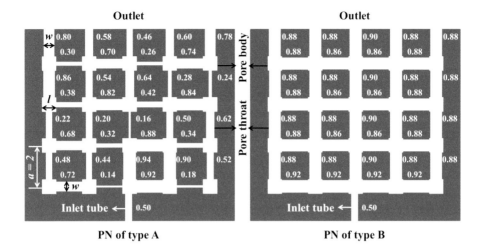

FIGURE 7.6 Structures of the pore networks used in the two-phase flow experiment. The numbers are pore throat widths (the unit is mm).

widths are 0.86, 0.88, 0.90, or 0.92 mm. Both PNs have a size of 4×4 pore throats. Figure 7.6 shows the structures of these two PNs, where the numbers are the pore throat widths (the unit is mm). The middle pore body at one side of the PN is connected to an inlet tube of 0.5 mm wide and of 8 mm long, through which the invading fluid is injected. Opposite to this inlet side is the outlet open to the environment. The other two sides are impermeable. The PN is initially filled with the displaced fluid (air). The invading fluid is then injected into the PN until the breakthrough moment. In the drainage experiment, the invading fluid is water with an advancing contact angle of about 67°. In the imbibition experiment, the invading fluid is the mixture of 20% v/v water and 80% v/v alcohol with an advancing contact angle of about 103°. The PN is placed horizontally on a base to eliminate the effects due to gravitational forces.

The invading fluid is injected into the pore network using a syringe pump (Harvard Apparatus, 11 Plus, USA). The flow rate is controlled to 0.1 µl/min so as to achieve a low capillary number ($Ca \sim 10^{-8}$). The capillary number is defined as $Ca = \mu v / \sigma$, where σ is the surface tension, and μ and v are the viscosity and velocity of the invading fluid, respectively. The movement of the invading fluid is recorded by a camera equipped with a macro lens (Nikon D810, Japan).

7.3.2 PORE NETWORK MODEL FOR TWO-PHASE FLOW IN POROUS MEDIA

To simulate the capillary force–dominated two-phase flow in a PN, the following algorithm is employed. At each step, only the available pore with the lowest threshold pressure is invaded. A pore is trapped if it is filled with the displaced fluid and there is no flow path between it and the outlet. A pore is available if it is filled with displaced fluid, not trapped, and adjacent to an invaded duct. The threshold pressure of an available pore is determined by Eqs. (7.2)–(7.7). If the capillary valve effect is

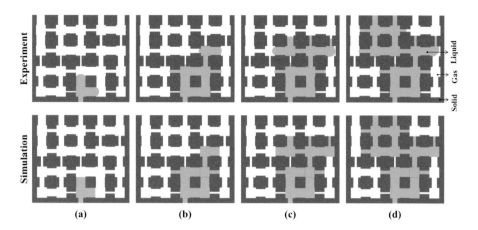

FIGURE 7.7 Comparison between drainage in the PN of type A obtained from the experiment and the PN model with the CVE. Figure 7.7(d) is the breakthrough moment.

neglected, the threshold pressures of all available ducts will be computed by Eq. (7.2) with w being the pore width.

Figures 7.7–7.9 compare the numerical and experimental results for the two-phase flows in the PNs of types A and B. The numerical results are obtained from the PN model with the CVE. In the numerical results, the pore body invaded through merging invasion is marked by a box. As can be seen from Figures 7.7–7.9, the numerical results agree well with the experimental observations. This validates the effectiveness of the developed PN model.

For the two-phase flow in the PN of type A, pore invasion is dominated by bursting invasion, Figures 7.7 and 7.8. Drainage and imbibition are similar. The only difference is that more pore throats are invaded in imbibition. In the imbibition case,

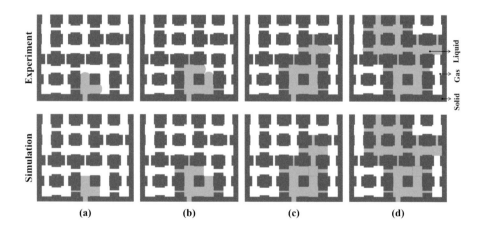

FIGURE 7.8 Comparison between imbibition in the PN of type A obtained from the experiment and the PN model with the CVE. Figure 7.8(d) is the breakthrough moment.

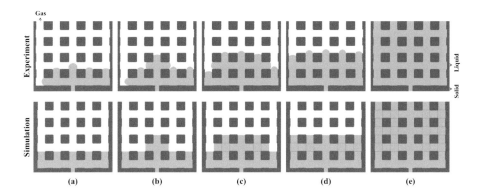

FIGURE 7.9 Comparison between drainage in the PN of type B obtained from the experiment and the PN model with the CVE. Figure 7.9(e) is the breakthrough moment.

pore throats have a lower threshold pressure than pore bodies; and all available pore throats are invaded before the next pore body invasion. In drainage, a pore throat could have a larger threshold pressure than a pore body; hence, not all available pore throats are invaded before invasion into a pore body. Both drainage and imbibition show the following invasion characteristics: before pore body invasion, the invading fluid has occupied the largest pore throat between an invaded pore body and a non-trapped pore body full of the displaced fluid; the available pore body connected to this largest invaded pore throat will be invaded in the next pore body invasion. As a result, drainage and imbibition in the network of type A are similar and exhibit a capillary fingering pattern, Figures 7.7 and 7.8.

For drainage in the PN of type B, pore invasion is dominated by merging invasion, see Figure 7.9. In this case, the pores have similar sizes; and the invasion processes are affected significantly by the fabrication uncertainty. Differences in the invasion order are found between the simulation and the experiment, e.g., invasion into the left down pore. The experimental results are expected to be the same as the numerical results if the fabrication uncertainty is reduced. For this reason, we focus on the numerical results in the following analysis.

The PN is divided into four layers from the inlet to the outlet. Each layer has five pore bodies and four pore throats. The pore throats in the first layer are larger than others, Figure 7.6. Hence this layer is invaded first. This resembles invasion from an open face of the PN. All pore throats between the first and second layers are invaded before invasion into a pore body in the second layer, Figure 7.9a. This is because threshold pressures for bursting invasion into pore bodies are larger than those of pore throats.

A pore body in the PN of type B will be invaded by merging invasion if two neighboring menisci are attached to this pore body. Merging invasion has a lower threshold pressure than bursting invasion. Hence, after the invading fluid occupies a pore body in the second layer and its connected pore throats, the available pore bodies in the second layer adjacent to this invaded pore body are invaded in the next step, Figure 7.9b and 7.9c. Pore body invasion in the third layer will not happen

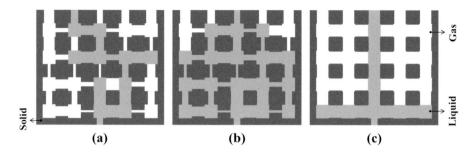

FIGURE 7.10 Phase distributions at the breakthrough moment obtained from the PN model without the CVE: (a) drainage in the PN type A; (b) imbibition in the PN of type A; (c) drainage in the PN of type B.

unless all the pore bodies in the second layer are invaded, Figure 7.9d. Invasion in the left network layers follows similarly. Thus, a stable invasion pattern is observed, Figure 7.9e.

Figure 7.10 shows the phase distributions at the breakthrough moment predicted by the PN model without the CVE. The numerical and experimental results are similar for drainage in the PN of type A, Figures 7.7d and 7.10a. They are quite different for imbibition in the PN of type A, Figures 7.8d and 7.10b. The huge difference is also observed for drainage in the PN of type B, Figures 7.9e and 7.10c.

The results presented in the present section show clearly that the CVE should be considered in PN models for two-phase transport in porous media.

7.4 CAPILLARY VALVE EFFECT ON SLOW DRYING IN POROUS MEDIA

Based on the PN model with the CVE developed in Section 7.3 for capillary force–dominated two-phase flows in porous media, we introduce the PN model with CVE for slow drying in porous media (Wu et al., 2016b) in this section. In the model, the vapor diffusion is considered. To understand in detail the influence of CVE, we consider only the slow drying case and neglect the effects of gravity, viscosity, liquid film flow, and heat transfer. To validate the developed model, a drying experiment with a PDMS microfluidic PN is performed.

7.4.1 SLOW DRYING EXPERIMENT

The slow drying experiment is performed with a microfluidic PN. Since effects of gravity, viscosity, and liquid films in corners of pores in the PN are not considered in the developed PN model, these effects need to be eliminated in the experiment. To suppress effects of corner films, a PDMS microfluidic PN is used. This is because PDMS has not only a transparent nature but also a hydrophobic surface characteristic. During drying of a PDMS PN filled with water, the angle between the moving gas-liquid interface and the liquid-solid interface is about 69°. This means that corner films cannot exist in the PN pores.

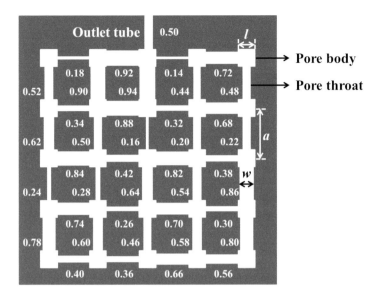

FIGURE 7.11 Structure of the PN used in the drying experiment.

It should be noted that PDMS is permeable to water. To reduce water loss due to permeation, the PN is covered by a glass sheet 1 mm thick. We find that if the PN is not covered by the glass sheet, the total drying time will be about three times shorter. To eliminate the viscosity effect on the drying-induced two-phase flow in the PN, only one of the PN pores (outlet tube) is open to the environment through a long tube, Figure 7.11. The purpose of this long tube is twofold. First, it controls the drying rate at a low level, which guarantees that the two-phase flow in the PN is dominated by capillary forces and the viscosity effect can be neglected. Second, this tube gives an explicit boundary condition for the mass transfer between the environment and the PN, which can be easily incorporated in the PN model. To relieve the gravity effect, the network is placed horizontally during the drying experiment.

The microfluidic PN, schematically shown in Figure 7.11, has a similar structure as the one introduced in subsection 7.3.1. The numbers shown in Figure 7.11 are the pore throat width (the unit is mm). All the PN sides are closed except that the middle pore at one side of the PN is open to the environment through a tube of 0.5 mm wide and $\delta = 8$ mm long (for convenience, this tube is called the outlet tube). All the PN pores and the outlet tube have the same depth of $h = 0.1$ mm in the direction perpendicular to the plane shown in Figure 7.11.

Initially, the microfluidic PN is immersed into de-ionized water contained in a plastic cylinder depressurized by a vacuum pomp so as to saturate the PN with water. Then, the PN is placed horizontally on a black plate in a chamber with almost constant temperature ($23.8 \pm 1°C$) and relative humidity ($24 \pm 2\%$). Time variation of the liquid distribution in the network is recorded by a camera (NIKON D810, Japan) equipped with a macro lens (AF-S VR Micro-Nikkor 105 mm f/2.8G IF-ED, Japan). The camera is controlled by a computer through the software of Camera Control Pro 2, and images are acquired at a time interval of 5 min.

Captured images have a resolution of 7360×4912 pixels, which provides a spatial resolution of about 5 μm per pixel.

7.4.2 PORE NETWORK MODEL FOR SLOW DRYING IN POROUS MEDIA

To simulate the process of slow drying in the PN shown in Figure 7.11, it is necessary to determine the threshold pressure of each partially filled pore. For each liquid cluster in the PN, the partially filled pore with the lowest threshold pressure will be emptied first during drying (Wu et al., 2014). The threshold pressure of each pore is determined by Eqs. (7.2)–(7.7), for which $\sigma = 0.0728$ N/m is the surface tension, and $\theta_a = 69°$ is the advancing contact angle.

Vapor transport between two neighboring pores in the PN during drying is described as a one-dimensional steady diffusion through stagnant air. The diffusion rate (mol/s) from a pore body to one of its neighboring pore throats with width w is:

$$Q = \frac{\dfrac{2hP_gD}{RT}\ln\dfrac{P_g - P_{v,t}}{P_g - P_{v,p}}}{1 + l_t / w} \tag{7.15}$$

where $h = 0.1$ mm is the pore depth, $l_t = 1$ mm the pore throat length, $T = 23.8°C$ the system temperature, $R = 8.314$ J/(mol K) the gas constant, and D the diffusivity of vapor, determined as $D = 2.14 \times 10^{-5}[(T + 273)/273]^{2.87}$ m²/s (Lee and Wilke, 1954). In Eq. (7.15), $P_g = 1.014 \times 10^5$ Pa is the total gas pressure, and $P_{v,p}$, and $P_{v,t}$ are the vapor pressures in pore bodies and pore throats, respectively. The vapor pressure in a pore is represented by the one at the pore center; for a partially filled pore, the vapor pressure is equal to the saturated one, which is about 2.984×10^3 Pa at $T = 23.8°C$. Similarly, the diffusion rate between the environment and the PN pore attached to the outlet tube with width w is:

$$Q = \frac{\dfrac{2hP_gD}{RT}\ln\dfrac{P_g - P_{v,t}}{P_g - P_{v,p}}}{1 + 2\delta / w} \tag{7.16}$$

where δ is the length of the outlet tube and $P_{v,e}$ is the vapor pressure in the environment with the relative humidity of 24%.

Now, the slow drying process in the PN shown in Figure 7.11 can be simulated straightforwardly using the following algorithm: (1) identify liquid clusters in the PN, (2) determine the vapor pressure in each empty pore in the PN, (3) calculate the drying rate of each liquid cluster, (4) identify the invasion pore of each liquid cluster, i.e., the partially filled pore with the lowest threshold pressure, (5) calculate the time to empty the invasion pore of each liquid cluster as the volume of liquid in this pore divided by the drying rate of the cluster, and select the minimum time as the step time, (6) update the liquid saturation in the invasion pore of each liquid cluster based on the step time determined in step (6), and (7) repeat the previous steps until all liquid in the PN is removed.

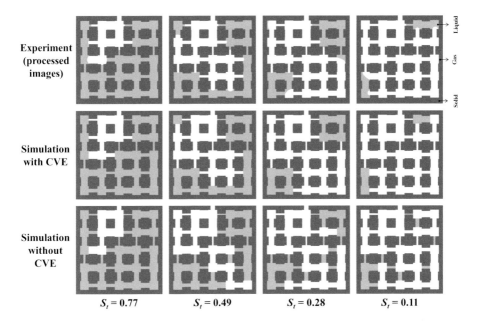

FIGURE 7.12 Variation of the liquid distribution in the network with the total liquid saturation S_t.

In previous PN models for drying porous media, however, the CVE is neglected, and the threshold pressures of a partially filled pore is determined by Eq. (7.2), for which w is the pore width (for a square pore body, the width is equal to the length).

The drying processes in the PN shown in Figure 7.11 (i.e., variations of the liquid distribution, number of liquid clusters, and liquid saturation) obtained from the experiment and the PN models with and without the CVE are compared in Figures 7.12–7.14. As can be seen, the numerical results have a better agreement with the experimental data if the CVE is considered in the PN model.

If the CVE is not considered in the model, the threshold pressure of a partially filled pore with width w and depth h is determined as $2\sigma\cos\theta_a(1/w + 1/h)$, which implies that for $0 < \theta_a < \pi/2$, pore bodies always have a lower threshold pressure than the connected pore throats since pore bodies are larger than the adjacent pore throats. Hence, once a pore throat in a liquid cluster is invaded by gas, its neighboring filled pore body will be emptied immediately in the next step, rendering a "throat-body" emptying pattern. This in turn results in many isolated filled pore throats in the PN (Figure 7.12) since pore throats are more numerous than pore bodies. In the model with the CVE, the threshold pressure of a pore body can be larger than that of a pore throat. For example, the threshold pressure for bursting invasion into a pore body from a pore throat with width 700 μm is greater than the threshold pressure of a pore throat with width 240 μm, according to Eqs. (7.2) and (7.3). As a result, the PN model without the CVE predicts more isolated filled throats (Figure 7.12) and more liquid clusters (Figure 7.13) as compared to the experiment and the model with the CVE.

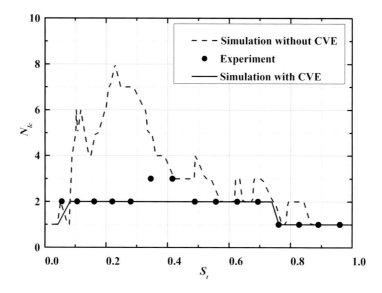

FIGURE 7.13 Variation of the number of liquid clusters, N_{lc}, with the total liquid saturation S_t.

As shown in Figure 7.14, variations of the liquid saturation with the normalized drying time (i.e., the drying time divided by the total drying time) obtained from the experiment and the PN simulations are similar. Nevertheless, the total drying times predicted by both the models with and without the CVE are about

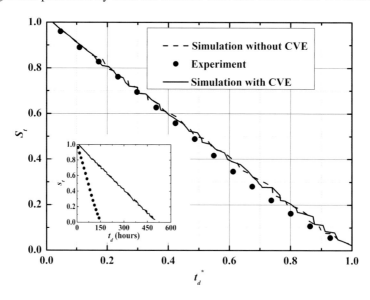

FIGURE 7.14 Variation of the total liquid saturation S_t with the normalized drying time t_d^*. The normalized drying time is the drying time t_d divided by the total drying time. The inset shows variation of the liquid saturation with the drying time.

511h, much longer than the experimental data of about 152h, see the inset of Figure 7.14. That is because both the liquid permeation through the PDMS network and the roughness film flow along the pore walls can accelerate the drying process (Pillai et al., 2009). These two factors, however, are not included in the present models.

The comparison between PN modeling and experimental results show clearly that the numerical results have a better agreement with the experimental data if the CVE is considered in the PN model.

7.5 CAPILLARY VALVE EFFECT ON CONVECTIVE DRYING IN POROUS MEDIA

Based on the PN model with the CVE developed in Section 7.4 for slow drying in porous media, we introduce the PN model with the CVE for convective drying in porous media (Wu et al., 2017) in this section. In the model, the liquid flow is taken into account. We take evaporation from the cathode gas diffusion layer (GDL) during gas purge as an example to study convective drying of porous media. GDLs used in the cathode of proton exchange membrane fuel cells are usually hydrophobic. The drying process in a GDL bounded with a gas purge channel is illustrated in Figure 7.15. Isothermal conditions are considered, and the environmental temperature is set to 293 K. The GDL is partially covered by the land of the bipolar plate. The gas channel (GC) has a width of $W_C = 0.5$ mm and a height of $H_C = 0.5$ mm. The width of the land is $W_L = 0.25$ mm. The GDL has a width of $W_G = 1$ mm and a height of $H_G = 0.25$ mm. Both the GDL and the GC have the same length of $L = 1$ mm. The GDL is initially filled with liquid water, and then dry purge gas is injected into the GC.

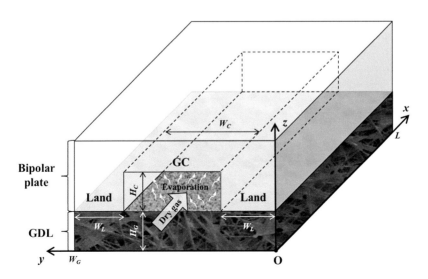

FIGURE 7.15 Schematic of drying of GDL bounded with a gas purge channel.

The gas flow in the GC is assumed as fully developed laminar flow. The gas flow is zero in the y- and z-directions, and the gas velocity u_c and flow rate Q_c along the x-direction are determined as (White, 1991)

$$u_C = \frac{4W_C^2}{\mu_g \pi^3} \frac{\Delta P}{L} \sum_{i=1,3,5}^{\infty} \left[1 - \frac{\cosh\left(i\pi \dfrac{z - H_G - \dfrac{H_C}{2}}{W_C} \right)}{\cosh\left(i\pi \dfrac{\dfrac{H_C}{2}}{W_C} \right)} \frac{\cos\left(i\pi \dfrac{y - W_L - \dfrac{W_C}{2}}{W_C} \right)}{i^3} \right]$$ (7.17a)

$$Q_c = \frac{H_C W_C^3 \rho_g}{12\mu_g} \frac{\Delta P}{L} \left[1 - \frac{192}{\pi^2} \frac{W_C}{H_C} \sum_{i=1,3,5}^{\infty} \frac{\tanh\left(i\pi \dfrac{H_C}{2W_C} \right)}{i^5} \right]$$ (7.17b)

where μ_g is the dynamic viscosity of gas, and $\Delta P/L$ is the pressure drop along the channel. The Reynolds number of the gas flow in the channel is defined as:

$$Re = \frac{Q_c D_c \rho_g}{A_c \mu_g}$$ (7.18)

where A_C is the cross-sectional area of the gas channel, D_C the hydraulic diameter, and ρ_g the gas density. The vapor transfer in the GC is described by:

$$u_c \frac{\partial C}{\partial x} = D\left(\frac{\partial^2 C}{\partial x^2} + \frac{\partial^2 C}{\partial y^2} + \frac{\partial^2 C}{\partial z^2} \right)$$ (7.19)

where C is the molar concentration of vapor and D the diffusivity of vapor in the gas phase.

The gas flow from the GC to the GDL is not considered here since the gas velocity is much smaller in the GDL than in the GC. For $Re = 100$ used here, the averaged gas velocity in the GC is $u_{C,m} = 3.14$ m/s, and the pressure drop along the GC is 6.65 Pa. If the pressure drop along the x-direction in the GDL is the same as that in the GC, the average gas velocity in the GDL of an absolute permeability of 5×10^{-12} m² (typical value for carbon paper GDL) would be about $u_{G,m} = 1.79 \times 10^{-3}$ m/s, according to Darcy's law. Obviously, the gas flow velocity is much smaller in the GDL than in the GC. For this reason, the gas flow from the GC to GDL is neglected.

Because of the low gas velocity in the GDL, the Péclet number for vapor transfer in the GDL is very low, about 1.39×10^{-3}. The Péclet number represents the relative importance of convection and diffusion for the mass transfer and is defined as $Pe = (u_{G,m} \cdot d_p)/D$, where $d_p = 20$ μm is the average pore size of the GDL. For such a low Péclet number, the vapor transport in the GDL is dominated by diffusion and the convective transport can be neglected.

7.5.1 Pore Network Model for Convective Drying in Porous Media

The void space of a GDL is approximated by a PN composed of cubic pore bodies connected by cylindrical pore throats. The cylindrical pore throats are used since the threshold pressure for bursting invasion into pore bodies from cylindrical pore throats can be theoretically obtained, see Section 7.2. A two-dimensional scheme of the generated PN is shown in Figure 7.16. The size of a pore is defined by the radius of the largest inscribed sphere. The radii of pore bodies and throats are uniformly distributed in the range of [8–12] and [2–7] μm, respectively. The distance between the centers of two neighboring pore bodies is $a = 25$ μm. A pore body and its connected pore throats yield a cubic unit cell, which has a side length of 25 μm. The number of pore bodies (or unit cells) is 10 in the z-direction and 40 in the x- and y-directions. The generated PN has a porosity of about 0.6 and an absolute permeability of 4.5×10^{-12} m².

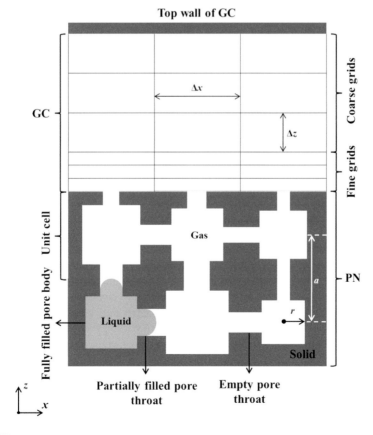

FIGURE 7.16   Two-dimensional view of the generated 3D PN bounded with a GC.

The vapor transport in the PN is described by Fick's law, and the diffusion rate from pore body i to pore throat j is determined as:

$$Q_{d,ij} = g_{d,ij}\left(C_i - C_j\right) \tag{7.20a}$$

$$\frac{1}{g_{d,ij}} = \frac{1}{g_{d,i}} + \frac{1}{g_{d,j}} \tag{7.20b}$$

$$g_{d,i} = D\frac{(2r_i)^2}{l_i/2}, \quad g_{d,j} = D\frac{\pi r_j^2}{l_j/2} \tag{7.20c}$$

where $Q_{d,ij}$ is the diffusion rate from pores i to j, g_d the diffusion conductivity, and r and l the pore radius and length, respectively. The liquid flow in the PN is considered as fully developed laminar flow, and the flow rate from pore body i to pore throat j is determined as:

$$Q_{f,ij} = g_{f,ij}\left(P_{l,i} - P_{l,j}\right) \tag{7.21a}$$

$$\frac{1}{g_{f,ij}} = \frac{1}{g_{f,i}} + \frac{1}{g_{f,j}} \tag{7.21b}$$

$$g_{f,i} = \frac{\rho_l}{M_l}\frac{\pi r_i^4}{8\mu_l(l_i/2)}, \quad g_{f,j} = \frac{\rho_l}{M_l}\frac{\pi r_j^4}{8\mu_l(l_j/2)} \tag{7.21c}$$

where $Q_{f,ij}$ represents the flow rate from pores i to j, g_f the flow conductivity, P_l is the liquid pressure, μ_l the dynamic viscosity of liquid, ρ_l the liquid density, and M_l the molecular weight of liquid.

Drying of the generated PN is related to movement of menisci therein, which in turn depends on the threshold pressure of each pore in the PN. The threshold pressure of each pore in the PN was introduced in Section 7.2. The advancing contact angle is taken as 105° since we assume that the PN is hydrophobic.

The vapor concentration in the GC, described by Eq. (7.19), is computed using the finite volume method. The space of the GC is divided into a number of regular grid elements, the size of which is designed so that each grid element at the PN – GC interface contacts only one unit cell in the PN, Figure 7.16. The number of grid elements is $N_x = 40$ in the x-direction and $N_y = 20$ in the y-direction, and all grid elements have the same side length in these two directions, i.e., $\Delta x = \Delta y = 25$ μm. The number of grid elements in the z-direction is $N_z = 20$. The grid elements along this direction are divided equally into two groups of fine and coarse grid, Figure 7.16. The fine grid near the PN – GC interface has $\Delta z = 5$ μm, whereas $\Delta z = 20$ μm is used for the coarse grid. The grid resolution is adequate based on the grid independent test.

Based on the generated grids in the GC, Eq. (7.19) is discretized. For vapor transport in the PN, the mass conservation law is applied to each empty pore, i.e., the sum of the diffusion rates to each empty pore is zero. In this way, linear equations for vapor concentration in the GC and the PN are obtained. All linear equations are solved using the Gauss-Seidel iteration method. The iteration is stopped when the maximal relative change between two successively obtained values is smaller than 10^{-5}.

The boundary conditions for vapor transport in the GC and the PN are as follows: At the inlet of the GC, the vapor concentration is zero; at the outlet of the GC, the gradient of vapor concentration is zero along the x-direction. All the GC walls and the sides of the PN are impermeable except the PN–GC interface, where the vapor concentration follows:

$$D \frac{\pi r_t^2}{l_t / 2}(C_t - C_{int}) = D \frac{\Delta x \Delta y}{\Delta z / 2}(C_{int} - C_g) \tag{7.22}$$

where C_{int} is the vapor concentration at the PN–GC interface, and C_t and C_g are the vapor concentrations in the PN pore and the GC grid element near the interface, respectively. In the PN, the vapor concentration of pores containing liquid is equal to the saturated vapor concentration, which is 1.21×10^{-2} mol/m³.

The vapor concentrations in the GC and PN can be computed using the following algorithm:

1. Vapor at the PN–GC interface is assumed to be the saturated.
2. The linear equations for vapor concentrations in the GC and PN are solved.
3. The vapor concentrations at the PN–GC interface are updated based on Eq. (7.22).
4. Steps (2) and (3) are repeated until the maximal relative change between any two successively obtained vapor concentrations at the PN–GC interface is smaller than 10^{-5}.

During drying of the PN, three types of pores can be discerned, as illustrated in Figure 7.16. An empty pore is filled with gas. A partially filled pore contains liquid with at least one attached meniscus. A fully filled pore is full of liquid without empty space adjacent to it. The partially and fully filled pores are also called filled pores. Liquid films are not considered because of the hydrophobic nature of the PN.

The liquid pressures of filled pores in the PN are gained based on the mass conservation law. For each fully filled pore, the sum of flow rates into this pore is zero. For each partially filled pore with static menisci, the sum of liquid flow rate into this pore is equal to the sum of vapor diffusion rates out of this pore. For each partially filled pore with moving menisci, the liquid pressure is equal to the difference between the gas pressure and the threshold pressure of the pore, i.e., $P_l = P_g - P_{th}$, for which $P_g = 1.01 \times 10^5$ Pa is the gas pressure. The sides of the PN are impermeable to liquid flow.

The liquid pressure in the PN is gained using the following algorithm:

1. The liquid clusters in the PN are identified.
2. All menisci in the PN are assumed to be static.

3. The partially filled pores with the lowest threshold pressure are determined for each liquid cluster, and the menisci attached to these pores are set to be moving.
4. The pressure field in the PN is solved based on the mass conservation law.
5. In each liquid cluster, the value of $P_g - P_{th} - P_l$ is determined for each static meniscus, and the meniscus with the largest positive value is set to be moving.
6. Steps (3)–(5) are repeated until no meniscus changes from static to moving.

Based on the liquid and vapor transport in the PN mentioned above, the drying process in the PN is now can be simulated straightforwardly using the following procedure:

1. The liquid clusters in the PN are identified.
2. The vapor concentration fields in the GC and the PN are solved.
3. The liquid pressure field and the moving menisci in the PN are determined.
4. The liquid removal rate from each partially filled pore with moving menisci is calculated as the difference between the rate of vapor diffusion out of this pore and the rate of liquid flow into this pore.
5. The time to empty each partially filled pore with moving menisci is computed, and the minimum one is selected as the time step.
6. Volumes of liquid in the partially filled pores with moving menisci are updated based on the time step determined in step (5).
7. The above steps are repeated until the preset condition is satisfied.

7.5.2 Simulation Results for Convective Drying in Porous Media

The generated PN is initially filled with liquid water; then dry gas is injected into the GC. The Reynolds number of the gas flow is set as $Re = 100$. Drying-induced gas invasion into the pore bodies is dominated by bursting invasion since merging invasion is impossible according to the pore size distribution and wettability of the generated PN. Our simulation results indicate that the capillary forces dominate over the viscous forces, and the gas invasion process depends mainly on the threshold pressures of pores.

Owing to the hydrophobic nature of the PN, the pore bodies have a larger threshold than the pore throats. To this end, once a pore body is invaded during drying, its neighboring throats will be emptied in the next steps. In a liquid cluster, the partially filled pore with the lowest threshold pressure will be invaded. The threshold pressure of a pore body depends on the sizes of its neighboring pore throats. The sizes of pore throats are randomly distributed. This in turn results in random gas invasion in the PN, see Figure 7.17. In this figure, variation of liquid distribution in the PN during drying is presented. The PN and GC are also sketched. The filled pore bodies and throats in the PN are shown by spheres and lines, respectively; the empty pores and the solid matrix are not shown for clarity. S_t denotes the total liquid saturation in the PN. As shown in Figure 7.17, liquid seems to be randomly distributed in the PN during drying, indicating an inherently random gas invasion process.

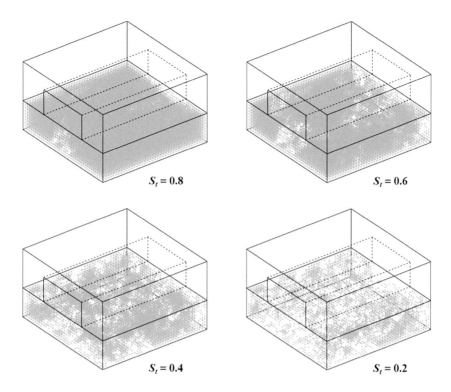

$S_t = 0.8$

$S_t = 0.6$

$S_t = 0.4$

$S_t = 0.2$

FIGURE 7.17   Variation of the liquid distribution in the PN during drying. S_t denotes the total liquid saturation in the PN.

Variations of the liquid saturations under the GC (S_C) and the land (S_L) during drying are compared in Figure 7.18. The liquid saturation under the GC (or land) is defined as the ratio of the liquid volume in the PN under the GC (or land) to the total volume of pores in the PN. At the beginning of drying (i.e., $S_t > 0.9$), owing to the screen effect of the land (i.e., PN under land is not available to gas), gas invades mainly the PN pores under the GC and the liquid saturation under the land remains almost constant. This in turn results in a lower liquid saturation under the GC than under the land. As the total liquid saturation varies from $S_t = 0.8$ to 0.1, the liquid saturation under the GC reduces from 0.33 to 0.003; the liquid saturation under the land decreases from 0.47 to 0.1. The decrease rates of S_C and S_L are similar. This further indicates that gas invasion in the PN is an inherently random process.

Variation of the ratio of the drying rate from the zone under the GC to the total drying rate, denoted by f, is also shown in Figure 7.18. When the total liquid saturation is $S_t \geq 0.3$, the drying rate from the zone under the GC is more than 90% of the total drying rate. This means that drying rate from the zone under the GC is much higher than that under the land. If no liquid flow paths exist between the zones under the land and the GC, liquid under the GC will be removed first, and then liquid under the land is removed. Such a drying pattern, i.e., through-plane drying under the GC followed by in-plane drying under the land, has been employed in previous

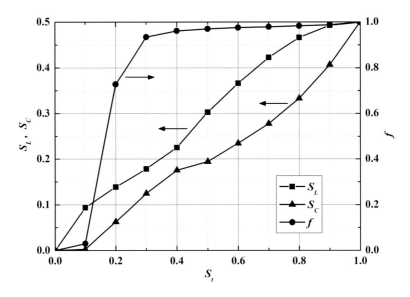

FIGURE 7.18 Variation with the total liquid saturation of the liquid saturations under the GC and the land as well as of the fraction of the drying rate from the zone under the GC. The liquid saturation under the land and the GC is denoted by S_L and S_C, respectively. S_t is the total liquid saturation. The ratio of the drying rate from the zone under the GC to the total drying rate is denoted by f.

theoretical models to study GDL drying (Sinha and Wang, 2007). This drying pattern, however, is not applicable when liquid under the land and the GC is well connected; in this case, drying-induced gas invasion in the PN is a random process.

The total drying rate from the PN, E_d, versus the total liquid saturation is shown in Figure 7.19. In this figure, the total drying rate is normalized to the initial drying rate. Three drying regimes can be distinguished: a surface evaporation period, a constant rate period, and a falling rate period, consistent with the experimental findings (Cho and Mench, 2010). Since the PN is hydrophobic, all the pore throats at the PN – GC interface will be emptied first before gas invades the pore bodies, which in turn results in a sharp decrease in the total drying rate at the very beginning of drying. Then all the pore bodies adjacent to the PN – GC interface become partially filled, and those with a low threshold pressure will be invaded. After this surface evaporation period, gas invades mainly the pores inside the PN, and the liquid loss by evaporation from the pores near the PN – GC interface is compensated by the liquid capillary flow, thereby resulting in a slow decrease in the total drying rate (i.e., the so called constant rate period). After this, the total drying rate is significantly reduced, leading to the falling rate period.

We also present in Figure 7.19 the variation of the total area for vapor transfer (A_{vt}) between partially filled pores and their neighboring empty pores. Here, A_{vt} is normalized to the cross-sectional area of the PN. For a partially filled pore throat, the area for vapor transfer is the cross-sectional area of this pore throat multiplied by the number of neighboring empty pore bodies. For a partially filled pore body, the

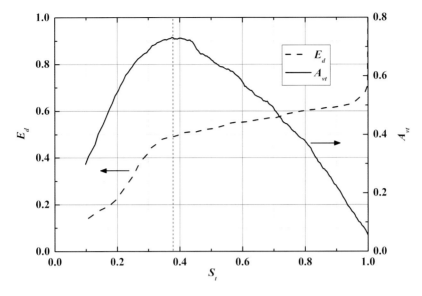

FIGURE 7.19 Variation with the total liquid saturation of the total drying rate as well as the total area for vapor transport between partially filled pores and their neighboring empty pores. The total drying rate is normalized to the initial drying rate. The total area for vapor transport is normalized to the cross-sectional area of the PN.

area for vapor transfer is the sum of the cross-sectional areas of the connected empty pore throats. The value of A_{vt} is proportional to the number of the menisci.

During drying of the PN, the value of A_{vt} increases first but then decreases, see Figure 7.19. The total liquid saturation at which the value of A_v is the maximum, i.e., $S_{t,A}$, is marked by the vertical dashed line. Although the transition between the constant rate and falling rate periods is smooth and it is not easy to define exactly the critical total liquid saturation for this transition, $S_{t,c}$, we still can see from Figure 7.19 that $S_{t,c}$ is close to $S_{t,A}$. This result is also observed in the drying cases of various PN thicknesses and boundary conditions. This indicates that there is an intrinsic correlation between $S_{t,c}$ and $S_{t,A}$.

As mentioned before, gas invasion into the pore bodies is dominated by bursting invasion for the investigated PN. In addition to bursting invasion, pore bodies can also be invaded by merging invasion. To also understand the characteristics of drying of a PN with dominant merging invasion, we construct a two-dimensional PN of size 3 × 3. We use such a small 2D PN in order to visualize clearly the gas invasion process. The structure of the PN is similar to that described in subsection 7.5.1 except that all the pore bodies have the same radius of 10 μm, and the radii of the pore throats are randomly distributed in the range [9.2–10] μm. One side of the PN is open to the environment, while the other sides are sealed. The vapor diffusion rate from a pore throat i at the open side of the PN to the environment is $\frac{D\pi r_i^2(C_i-C_e)}{\beta}$. Here, $C_e = 0$ is the concentration of vapor in the environment, and $\beta = 12.5$ μm is the diffusion distance between the open side of

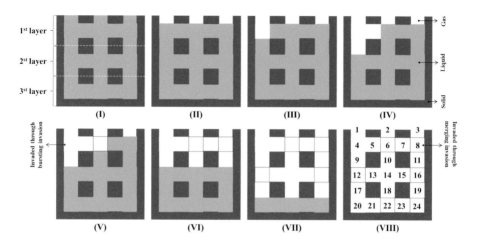

FIGURE 7.20 Variation of the liquid distribution in the PN during drying with dominant merging invasion.

the PN and the environment. To realize dominant merging invasion, all the pores in the PN have a moving contact angle of $\theta_m = 92°$.

Variation of liquid distribution in the PN during drying with dominant merging invasion is shown in Figure 7.20. The boxes are the pore bodies invaded through merging invasion. As shown in Figure 7.20, drying-induced gas invasion in the PN is a stable process. For this type of PN, pore throats have a lower threshold pressure than pore bodies; if a pore body has two empty pore throats neighboring (not opposite) to each other, this pore body will be invaded through merging invasion. For convenience of analysis, the PN is divided into three layers from the open side to its opposite side, each of which contains three pore bodies, Figure 7.20(I). The first layer is adjacent to the open side of PN. The pore throats and bodies in the PN are marked by the numbers shown in Figure 7.20(VIII).

For drying of the PN shown in Figure 7.20, the pore throats at the open side of PN are invaded first, leading all the pore bodies in the first layer to be partially filled, Figure 7.20(II). Then, the partially filled pore in the first layer with the lowest threshold pressure, i.e., pore body 4, is invaded through bursting invasion, Figure 7.20(III). Subsequently, the pore throats attached to pore body 4 are invaded, Figure 7.20(IV), since pore throats have a lower threshold pressure than pore bodies. After this, a new partially filled pore body 12 in the second layer is created, which has only one empty connected pore throat. By contrast, pore body 6, neighboring to pore body 4, in the first layer has two empty pore throats neighboring to each other.

Since the threshold pressure for merging invasion is lower than that for bursting invasion, pore body 6 in the first layer will be invaded through merging invasion before bursting invasion into the partially filled pore bodies in the second layer, Figure 7.20(V). Note that before gas invades the pore bodies in the second layer, all the pore bodies in this layer have at most one connected empty pore throat. To this end, gas will not invade the pore bodies in the second layer unless all the pore bodies in the first layer are invaded, Figure 7.20(VI). Gas invasion in the remaining layers of the PN is similar to that in the first layer, which in turn results in a stable gas invasion pattern, Figure 7.20(VII) and 7.21(VIII). As shown in Figure 7.20(VIII), only one pore body is invaded through bursting invasion in each layer.

During drying of a porous material with a stably receding drying front, as that shown in Figure 7.20, the drying rate is first sharply reduced and then continues with a relative rapid decrease as the total liquid saturation decreases; a period with nearly constant drying rate, as the one shown in Figure 7.19, cannot be observed in this case, see Wu et al. (2014). It can be concluded that the characteristics of drying of hydrophobic porous media depend significantly on the pore invasion pattern (bursting or merging invasion). The pattern of drying-induced gas invasion into pores in a porous material relies on the structure and wettability of this porous material. This explains why different drying characteristics were observed in literature for drying of hydrophobic porous media with different pore structures (Shahidzadeh-Bonn et al., 2007; Shokri et al. 2009; Cho and Mench, 2010).

7.6   SUMMARY

The pore network (PN) model for two-phase transport, especially drying, in porous media has been developed that takes into account the capillary valve effect (CVE). This effect is induced by the sudden geometrical expansion in the pore space. The PN modeling results have a better agreement with the experimental data if the CVE is considered in the model. This shows clearly that the CVE should be considered in the PN model so as to predict more accurately the two-phase transport in porous media. Because of the CVE, two types of invasion into pore bodies can be found: bursting and merging invasion. For drying of a porous material, if invasion into pore bodies is dominated by merging invasion, then the drying front is stably receded. But, when bursting invasion dominates invasion into pore bodies, the drying front is randomly distributed; and a nearly constant drying rate period can be observed. The characteristics of drying in porous media depend significantly on the pattern of invasion into pore bodies. However, in the present PN models, the corner films in porous media are not considered. Development of the PN models that take into account both the CVE and corner films is envisioned as a future study.

ACKNOWLEDGEMENTS

The first author is grateful for support of the National Natural Science Foundation of China (Nos. 51776122 and 51306124) and the Shanghai Pujiang Program (No. 17PJ1404600).

REFERENCES

Araujo, A.D., Vasconcelos, T.F., Moreira, A.A., Lucena, L.S. Andrade, J.S. 2005. Invasion percolation between two sites. *Physical Review E* 72: 041404.
Bazylak, A., Berejnov, V., Markicevic, B., Djilali, S.N. 2008. Numerical and microfluidic pore networks: Towards designs for directed water transport in GDLs. *Electrochimic Atca* 53: 7630–7637.

Blunt, M., King, M.J., Scher, H. 1992. Simulation and theory of two-phase flow in porous media. *Physical Review A* 46: 7680–7699.

Blunt, M. 2001. Flow in porous media – pore-network models and multiphase flow. *Current Opinion in Colloid & Interface Science* 6: 197–207.

Ceballos, L., Prat, M. 2013. Slow invasion of a fluid from multiple inlet sources in a thin porous layer: Influence of trapping and wettability. *Physical Review E* 87: 043005.

Chen, J.M., Huang, P.C., Lin M.G. 2008. Analysis and experiment of capillary valve for microfluidics on a rotating disk. *Microfluidics and Nanofluidics* 4: 427–437.

Cho, H., Kim, H.Y., Kang, J.Y., Kim, T.S. 2007. How the capillary burst microvalve works. *Journal of Colloid and Interface Science* 306: 379–385.

Cho, K.T., Mench, M.M. 2010. Fundamental characterization of evaporative water removal from fuel cell diffusion media. *Journal of Power Sources* 195: 3858–3869.

Duffy, D.C., Gillis, H.L., Sheppard, N.F., Kellogg, G.J. 1999. Microfabricated centrifugal microfluidic systems: Characterization and multiple enzymatic assays. *Analytical Chemistry* 71: 4669–4678.

Joekar-Niasar, V., Hassanizadeh, S.M. 2012. Analysis of fundamentals of two-phase flow in porous media using dynamic pore network models: a review. *Critical Reviews in Environmental Science and Technology* 42: 1895–1976.

Kharaghani, A., Metzger, T., Tsotsas, E. 2011. A proposal for discrete modeling of mechanical effects during drying, combining pore networks with DEM. *AIChE Journal* 57: 872–885.

Kharaghani, A., Metzger, T., Tsotsas, E. 2012. An irregular pore network model for convective drying and resulting damage of particle aggregates. *Chemical Engineering Science* 75: 267–278.

Knackstedt, M.A., Sheppard, A.P., Pinczewski, W.V. 1998. Simulation of mercury porosimetry on correlated grids: evidence for extended correlated heterogeneity at the pore scale in rocks. *Physical Review E* 58: R6923–R6926.

Lee, Y.C., Wilke, C.R. 1954. Measurement of vapor diffusion coefficient. *Industrial and Engineering Chemistry* 46: 2381–2387.

Lopez, R.H., Vidales, A.M., Zgrablich, G. 2003. Fractal properties of correlated invasion percolation patterns. *Physica A: Statistical Mechanics and its Application* 327: 76–81.

Mani, V., Mohanty, M.M. 1999. Effect of pore space spatial correlations on two-phase flow in porous media. *Journal of Petroleum Science and Engineering* 23: 173–188.

Metzger, T., Irawan, A., Tsotsas, E. 2007. Influence of pore structure on drying kinetics: a pore network study. *AIChE Journal* 53: 3029–3041.

Metzger, T., Irawan, A., Tsotsas, E. 2008. Viscous stabilization of drying front: three-dimensional pore network simulations. *Chemical Engineering Research and Design* 86: 739–744.

Moore, J.L., Cuiston, A. M., Mittendorf, I., Ottway, R., Johnson, R.D. 2011. Behavior of capillary valves in centrifugal microfluidic devices prepared by three-dimensional printing. *Microfluidics and Nanofluidics* 10: 877–888.

Pillai, K.M., Prat, M., Marcoux M. 2009. A study on slow evaporation of liquids in a dual porosity medium using square network model. *International Journal of Heat and Mass Transfer* 52: 1642–1656.

Plourde, F., Prat, M. 2003. Pore network simulations of drying of capillary porous media. Influence of thermal gradients. *International Journal of Heat and Mass Transfer* 46: 1293–1307.

Shahidzadeh-Bonn, N., Azouni, A., Coussot, P. 2007. Effect of wetting properties on the kinetics of drying of porous media. *Journal of Physics: Condensed Matter* 19: 112101.

Shokri, N., Lehmann, P., Or, D. 2009. Characteristics of evaporation from partially wettable porous media. *Water Resources Research* 45: W02415.

Sinha, P.K., Wang, C.Y. 2007. Gas purge in a polymer electrolyte fuel cell. *Journal of Electrochemical Society* 154: B1158–B1166.

Surasani, V.K., Metzger, T., Tsotsas, E. 2008. Consideration of heat transfer in pore network modeling of convective drying. *International Journal of Heat and Mass Transfer* 51: 2506–2518.

Taleghani, S.T., Dadvar M. 2014. Two dimensional pore network modeling and simulation of non-isothermal drying by the inclusion of viscous effects. *International Journal of Multiphase Flow* 62: 37–44.

Wilkiinson, D., Willemsen, J.F. 1983. Invasion percolation: a new form of percolation theory. *Journal of Physics A: Mathematical and General* 16: 3365–3376.

Vorhauer, N., Wang, Y.J., Kharaghani, A., Tsotsas, E., Prat, M. 2015. Drying with formation of capillary rings in a model porous medium. *Transport in Porous Media* 110: 197–223.

White, F.M. 1991. *Viscous fluid flow*, New York: McGraw-Hill.

Wu, R., Cui, G.M., Chen, R. 2014. Pore network study of slow evaporation in hydrophobic porous media. *International Journal of Heat and Mass Transfer* 68: 310–323.

Wu, R., Kharaghani, A., Tsotsas, E. 2016a. Capillary valve effect during slow drying of porous media. *International Journal of Heat and Mass Transfer* 94: 81–86.

Wu, R., Kharaghani, A., Tsotsas, E. 2016b. Two-phase flow with capillary valve effect in porous media. *Chemical Engineering Science* 139: 241–248.

Wu, R., Zhao, C.Y., Tsotsas, E., Kharaghani, A. 2017. Convective drying in thin hydrophobic porous media. *International Journal of Heat and Mass Transfer* 112: 630–642.

Yiotis, A.G., Boudouvis, A.G., Stubos, A.K., Tsimpanogiannis, I.N., Yortsos Y.C. 2004. Effect of liquid films on the drying of porous media. *AIChE Journal* 50: 2721–2737.

8 Taylor-series Expansion Method of Moments for Size-distributed Micro- and Nanoparticle Systems under Drying Condition

Mingzhou Yu
China Jiliang University, Hangzhou, Zhejiang, China

Yueyan Liu
China Jiliang University, Hangzhou, Zhejiang, China

CONTENTS

8.1 INTRODUCTION

Drying processes for size-distributed nanoparticles are found in a wide variety of fields such as chemical, agricultural, food, polymer, pharmaceutical, ceramics and aerosol-related environmental issues (Huang et al., 2017; Keshani et al., 2015;

Ramkrishna & Singh, 2014; Tsotsas & Mujumdar, 2014). Under drying condition, the heat and mass transfer of such systems is affected by external processes including convection and diffusion and internal processes such as nucleation, agglomeration or coagulation, breakage and surface growth (Friedlander, 2000). The key properties of such size-distributed particle systems is not only the interaction between particle phase and continuum phase but also the interaction among individual particles. The appropriate mathematical method for quantifying dynamical process is required to capture the size-distributed effect on fine particle dynamics. In the field of multiphase flows, however, the size of all particles in a system is usually assumed to be the same for simplicity. The mathematical models using the size assumption fail in the study on size-distributed nanoparticle two-phase system (Balachandar & Eaton, 2010).

Except for convection and diffusion transport, the evolution of micro- and nanoparticle dynamics arises mainly from internal mechanisms including homogeneous or heterogeneous nucleation, condensation, coagulation and breakage (Friedlander, 2000; Yu & Lin, 2018). Among these internal mechanisms, coagulation occurs most commonly, but it is the most difficult to be treated in mathematics. This is mainly because the correlations among all particles have to be considered separately. Since the pioneering work of Smoluchowski in 1917 (Smoluchowski, 1917), the mean-field theory has been introduced in aerosol collision problems and has been the basis for numerous theoretical applications. For micro- and nanoparticles under drying condition, especially aerosols, some important phenomena such as self-preserving distribution and gelation or asymptotic behavior have been studied using numerical methods such as the method of moments (MOM), the sectional method (SM) and the Monte Carlo method (MCM). Here, the MOM is superior to others in that it requires the least computational cost as well being relatively simple to implement. However, these achievements are usually limited to studies on internal dynamics such as coagulation, surface growth, nucleation and breakage. In the area of drying science and technology, the application of the MOM for resolving engineering problems is limited.

Although the MOM has become a powerful tool for investigating micro- and nanoparticle processes in most drying cases, the closure of moment-based governing equations is not easy to achieve. The MOM was first introduced by Hulburt and Katz in 1964 into the study of micro- and nanoparticle dynamics (Hulburt & Katz, 1964). This method primarily involves implementing a transformation from the particle size distribution function space $\{n(v)\}$ to the space of moments $\{m_k\}$, where $n(v)$ is particle number density in terms of particle size v and m_k is its k-th moment. According to the closure scheme, the MOM can be divided into methods including the predefined size-distributed method (log MM) (Lee, Chen, & Gieseke, 1984; Williams, 1985), quadrature-based MOM (QBMOM) [QMOM and direct QMOM (DQMOM)] (Marchisio & Fox, 2005; McGraw, 1997), p-th order–polynomial MOM (Barrett & Jheeta, 1996), MOM with interpolative closure (MOMIC) (Frenklach, 2002), and Taylor-series expansion MOM (TEMOM) (Frenklach & Harris, 1987; Yu et al., 2008). The MOMs share a common feature that renders them practically feasible: all of them require a finite number of equations through implementing the corresponding approximated techniques, containing lower-order moments to be solved. According to our thorough review of relevant literature, the QMOM, especially its variants, i.e., the direct quadrature MOM (DQMOM), is the most widely used MOM today. The Gaussian quadrature MOM has become a very successful tool for studying polydisperse multiphase flows on the basis of meso-scale descriptions (Fox, 2012).

Similar to the QMOM, the TEMOM does not require a specific size distribution of micro- or nanoparticles; therefore, it might have a wider scope of application relative to pre-assumed size-distributed techniques such as log MM. The TEMOM shows merit in efficiency when resolving some physicochemical processes; the classic version of the TEMOM (i.e., classic TEMOM) introduced in 2008 requires only three ordinary differential equations (ODEs) to be solved for the first three moments (Yu et al., 2008). Although an improved version of the TEMOM, the generalized TEMOM (GTEMOM), involves greater computational cost than the classic version, its efficiency is ensured (Yu et al., 2015). In this chapter, the application of TEMOM in nanoparticle formation and subsequent growth will be presented.

8.2   MICRO- AND NANOPARTICLE DYNAMICS

Drying processes for micro- and nanoparticles find applications in a wide range of powder industries, chemical engineering and environmental issues. Examples include nanoparticle synthesis for noble metal catalyst, spray drying for industrial milk powder or pharmaceutical tablets, and aerosol formation and subsequent growth under varying air conditions. The particle dynamics of these examples might be different due to different conditions, however, a general route for particle dynamics involving gas-to-particle conversion and subsequent particle size growth exits, i.e., micro- and nanoparticles form via nucleation through a heat and mass transfer process between particle phase and continuum phase, and then the particle grows up or its morphology varies due to coagulation (agglomeration or aggregation), condensation (surface growth) or coalescence. Under drying condition, the size of growing particles might reduce due to particle breakage or evaporation. Thus, the evolution of micro- and nanoparticle dynamics include both forward and inverse processes, i.e., the particle birth from precursor vapor and subsequent growth, and the particle volume reduction or even death, some of which are shown in Figure 8.1. Noted here, the external processes, including turbulent convection and diffusion, are not included.

In mathematics, an appropriate modeling approach for quantifying the dynamics shown in Figure 8.1 is the concept of the population balance equation (PBE, also called the general dynamic equation in the field of nuclear science and technology), which is capable of tracing the evolution of the particle number density by establishing a number balance equation. The PBE represents number balance of particles of a specific state, usually particle volume, diameter, component or shape factor (fractal dimension number). Without loss of generality, here we give the PBE involving both

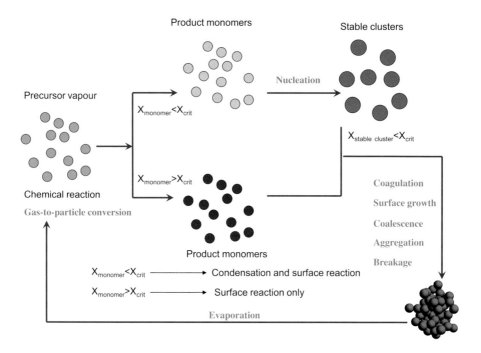

FIGURE 8.1 Micro- and nanoparticle dynamics under drying condition.

external and internal processes, in which all mechanisms affect the evolution of particle dynamics,

$$\frac{\partial n(v,\, x_i,\, t)}{\partial t} + \frac{\partial \big(u_j n(v,\, x_j,\, t) \big)}{\partial x_j} + \frac{\partial \big((u_{\mathrm{th}})_j n(v,\, x_j,\, t) \big)}{\partial x_j}$$

$$= \frac{\partial}{\partial x_j} \left(D_{\mathrm{B}} \frac{\partial n(v,\, x_j,\, t)}{\partial x_j} \right) + \frac{\partial \big(G_r n(v,\, x_i,\, t) \big)}{\partial v} + J\big(v^*,\, x_i,\, t \big) \delta\big(v - v^* \big)$$

$$+ \frac{1}{2} \int_{v^*}^{v} \beta(v - v',\, v') n(v - v',\, t) n(v',\, x_i,\, t)\, dv' \qquad (8.1)$$

$$- n(v,\, t) \int_{v^*}^{\infty} \beta(v,\, v') n(v',\, x_i,\, t)\, dv' + \int_{v}^{\infty} a(v') b(v|v') n(v',\, t)\, dv'$$

$$- a(v) n(v,\, t) + \dots,$$

where $n(v,\, x_i,\, t)$ is the particle number density for particle volume v, location x and time t; u_j is the particle velocity, u_{th} is the thermophoresis velocity, D_{B} is the Brownian diffusion coefficient, G_r is the particle surface growth rate, J is the nucleation rate for the critical monomer volume v^*, β is coagulation kernel between two particles and a and b are parameters accounting for the breakage rate associated with the turbulence shear force (Friedlander, 2000). Equation (8.1) encompasses almost all physicochemical processes of micro- and nanoparticles with a size smaller than approximately 1 μm, and therefore is reliable for studying fine particle dynamics under drying condition. In particular, an inherent advantage is that it can be coupled with the Navier–Stokes equation because it is a typical transport equation. However, the direct numerical solution of Equation (8.1) is intractable for most applications due to the extremely large number of independent variables, even if there is only one internal coordinate, and it should be further modified using suitable mathematical techniques.

In this chapter, the TEMOM will be introduced to solve Equation (8.1). A focus will be placed on the solution of Equation (8.1) involving coagulation.

8.2.1 The Closure Scheme of TEMOM

To quantify the evolution of particle number, it is necessary to define particle concentration as a function of time and particle volume. The disposition was first proposed by Smoluchowski (1917) for coagulation in dilute electrolytes, which has been a basis for solving micro- and nanoparticle multiphase problems. The integral form of Smoluchowski equation is (Müller, 1928):

$$\frac{\partial(v,\,t)}{\partial t} = \frac{1}{2}\int_0^v \beta(v-v',\,v')n(v-v',\,t)n(v',\,t)dv' - n(v,\,t)\int_0^\infty \beta(v,v')n(v',\,t)dv' \quad (8.2)$$

Equation (8.2) is actually a simplified form of Equation (8.1) regarding only coagulation. If the moment methodology is used, the general disposition for this problem is to convert Equation (8.2) into an ordinary differential equation with respect to moment m_k. The moment transformation involves multiplying Equation (8.2) by v^k and then integrating over the entire size distribution, and finally the converted moment equation is obtained (Yu et al., 2008):

$$\frac{dm_k}{dt} = \frac{1}{2}\int_0^\infty\int_0^\infty \kappa(v,v',\,k)n(v,\,t)n(v',\,t)dvdv' \quad (8.3)$$

where $\kappa(v,\,v',\,k) = \left[(v+v')^k - v^k - v'^k\right]\beta(v,\,v')$. The moment m_k is defined by

$$m_k = \int_0^\infty v^k n(v)dv \quad (8.4)$$

Some efforts have been made to achieve the closure of Equation (8.3). Five prominent techniques have been proposed by different researchers (Yu & Lin, 2018), i.e., making a prior assumption for the shape of the aerosol size distribution, approximating the integral moment by an n-point Gaussian quadrature, assuming the p-th order polynomial form for the moments, achieving closure with the interpolative method, and achieving closure with the Taylor-Expansion technique.

8.2.2 Classic Coagulation Theory

Most theoretical and experimental studies have shown Brownian coagulation to be the most common interparticle phenomenon for micro- and nanoparticles with diameter below 1 μm (Friedlander, 2000). The theory for resolving this issue is based on aerosol transport theory based on Stokes' law and Einstein's diffusion theory in the continuum regime, and kinetic theory of gases in the free molecular size regime (Pratsinis, 1988). In the transient regime between free molecular size regime and continuum regime, however, there is not an appropriate theory to solve this problem. The detailed definition for different regimes with respect to particle diameter can be found in the work of Pratsinis (1988) and Otto et al. (1999).

If the particle size falls into the free molecular size regime, then the collision frequency is obtained from the kinetic theory of gas:

$$\beta(v,v') = B_1 \left(\frac{1}{v} + \frac{1}{v'} \right)^{\frac{1}{2}} \left(v^{\frac{1}{3}} + v'^{\frac{1}{3}} \right)^2, \tag{8.5}$$

where $B_1 = (3/4\pi)^{1/6} (6k_bT/\rho)^{1/2}$, k_b is the Boltzmann constant, T is the gas temperature and ρ is the mass density of the particles. As Equation (8.5) is introduced into Equation (8.3) and then closed by the classic TEMOM, the set of moment equations can be finally written in the following form (Yu et al., 2008):

$$\begin{cases} \dfrac{dm_0}{dt} = \dfrac{\sqrt{2}B_1 \left\{ 65g^2 - 1210g - 9223 \right\}}{5184} m_1^{\frac{1}{6}} m_0^{\frac{11}{6}}, \\[4mm] \dfrac{dm_1}{dt} = 0, \\[4mm] \dfrac{dm_2}{dt} = -\dfrac{\sqrt{2}B_1 \left\{ 701g^2 - 4210g - 6859 \right\} g^{-\frac{1}{6}}}{2592} m_2^{\frac{1}{6}} m_1^{\frac{11}{6}}, \end{cases} \tag{8.6}$$

where $g = m_0 m_2 / m_1^2$. If the particle size approaches the mean free path of the gas and falls into the continuum-slip regime, the collision frequency can be obtained from the transport theory (Friedlander 2000):

$$\beta(v,v') = B_2 \left(\frac{C(v)}{v^{\frac{1}{3}}} + \frac{C(v')}{v'^{\frac{1}{3}}} \right) \left(v^{\frac{1}{3}} + v'^{\frac{1}{3}} \right), \tag{8.7}$$

where $B_2 = 2k_bT/3\mu$. The slip correction factor (v), is used to accommodate the gas slip effects for small particles.

$$C(v) = 1 + A\cdot Kn, \tag{8.8}$$

where $A = 1.591$, which is expected to make the collision kernel valid for the Knudsen number Kn up to about 5. Similar to the solution in the free molecular size regime, the set of moment equations disposed by Taylor-series expansion technique can be obtained in the following:

$$
\begin{cases}
\left.\dfrac{dm_0}{dt}\right|_{co} = B_2 \left\{ \dfrac{(-151m_1{}^4 + 2m_2{}^2 m_0{}^2 - 13m_2 m_1{}^2 m_0)m_0{}^2}{81m_1{}^4} \right. \\[2ex]
\qquad\qquad \left. + \dfrac{\hat{\lambda} m_0{}^{7/3}(5m_2{}^2 m_0{}^2 - 64 m_2 m_1{}^2 m_0 - 103 m_1{}^4)}{81 m_1{}^{13/3}} \right\} \\[3ex]
\left.\dfrac{dm_1}{dt}\right|_{co} = 0 \\[3ex]
\left.\dfrac{dm_2}{dt}\right|_{co} = B_2 \left\{ -\dfrac{2}{81} \dfrac{-151m_1{}^4 + 2m_2{}^2 m_0{}^2 - 13m_2 m_1{}^2 m_0}{m_1{}^2} \right. \\[2ex]
\qquad\qquad \left. - \dfrac{4}{81} \dfrac{\hat{\lambda} m_0{}^{1/3}(-2m_2 m_1{}^2 m_0 - 80 m_1{}^4 + m_2{}^2 m_0{}^2)}{m_1{}^{7/3}} \right\}
\end{cases}
\tag{8.9}
$$

where $\hat{\lambda} = A\lambda(4\pi/3)^{1/3}$.

Although Fuchs has proposed a general interpolation formula for collision frequency taking into account the transition from the free-molecule regime to the continuum regime (Fuchs, 1964), the Taylor-series expansion moment method cannot be applied due to the collision kernel's nonintegral form. Alternatively, we have to follow Otto's work and construct the TEMOM using the Dahneke's solution and the harmonic mean solution, respectively (Otto et al., 1999; Yu et al., 2011).

Following the study of Pratsinis (1988) and Park et al. (1999), the moment equation to cover the entire size regime by applying the harmonic mean method to TEMOMs takes the following form

$$
\left.\frac{dm_k}{dt}\right|_{entire} = \frac{dm_k/dt|_{co} \cdot dm_k/dt|_{fm}}{dm_k/dt|_{co} + dm_k/dt|_{fm}} \quad (k = 0,\, 1,\, 2)
\tag{8.10}
$$

In Otto's works, the moment equations using Dahneke's solution over the entire size regime is represented by

$$
\left.\frac{dm_k}{dt}\right|_{entire} = \left.\frac{dm_k}{dt}\right|_{co} \frac{1 + Kn_{m_k}}{1 + f(\sigma_g)Kn_{m_k} + 2Kn_{m_k}{}^2} \quad (k = 0,\, 1,\, 2)
\tag{8.11}
$$

with $Kn_{m_k} = \frac{1}{2}\left(\left.\frac{dm_k}{dt}\right|_{co}\right)\left(\left.\frac{dm_k}{dt}\right|_{fm}\right)^{-1}$.

Here $f(\sigma_g)$ is the correction function, which depends on the geometric standard deviation

$$
f(\sigma_g) = 2 + 0.7\ln^2\sigma_g + 0.85\ln^3\sigma_g
\tag{8.12}
$$

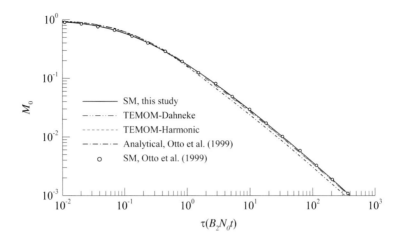

FIGURE 8.2 The comparison between TEMOM and the work by Otto et al. (1999) with respect to normalized particle number.

(*Source: J. Aerosol Sci.* 40 (2009) 549–562).

It should be noted that the geometric standard deviation in the TEMOM model is assumed to be the same with the log MM since log-normal size distribution is always considered to be reasonable in the evolution of aerosols (Pratsinis, 1988). The models given by Equation (8.10) and Equation (8.11) are named by TEMOM-Harmonic and TEMOM-Dahneke, respectively. Although Equation (8.12) was initially developed for spherical particles by Otto et al. (1999), it is found in our study that this equation can be extended to fractal-like aggregates in the TEMOM model without losing much accuracy. The TEMOM-Dahneke shows more accuracy than TEMOM-Harmonic, which is shown in Figure 8.2.

8.2.3 Closure Schemes

The key of any MOM is to solve moment ordinary differential equations using a limited number of known moments. For all unknown moments, there must be a suitable function that can approximate them by adjusting some parameters. In fact, for any MOM, knowing all the moments ($m_k, k = 0,1,\cdots,\infty$) or infinite moments provides a possibility to know the particle size distribution function using some reconstruction techniques (Frenklach, 2002). However, in practice, it is not possible to develop a MOM involving all the moments.

To obtain a suitable closure function for approximating any unknown moments, the internal coordinate of the PBE, such as particle volume v, must be replaced with its approximated polynomials using a Taylor-series expansion technique,

$$v^k = \left(\frac{u^{k-2}k^2}{2} - \frac{u^{k-2}k}{2} \right) v^2 + (-u^{k-1}k^2 + 2u^{k-1}k)v + u^k + \frac{u^k k^2}{2} - \frac{3u^k k}{2} + \cdots \quad (8.13)$$

In the original TEMOM (Yu et al., 2008), v^k was approximated using its three-order Taylor series, i.e.

$$v^k \approx \left(\frac{u^{k-2}k^2}{2} - \frac{u^{k-2}k}{2} \right)v^2 + (-u^{k-1}k^2 + 2u^{k-1}k)v + u^k + \frac{u^k k^2}{2} - \frac{3u^k k}{2} \quad (8.14)$$

As Equation (8.14) is introduced into Equation (8.4), the closure function for arbitrary moments can be easily obtained,

$$m_k = \int_0^\infty v^k n(v)dv = \left(\frac{u^{k-2}k^2}{2} - \frac{u^{k-2}k}{2} \right)m_2 + (-u^{k-1}k^2 + 2u^{k-1}k)m_1 + \left(u^k + \frac{u^k k^2}{2} - \frac{3u^k k}{2} \right)m_0 \quad (8.15)$$

m_k in Equation (8.15) can be considered to be a function of the first three moments m_0, m_1 and m_2. In principle, the number of moments in Equation (8.15) is consistent with the reserved terms of the Taylor series in Equation (8.6). For example, it is necessary to take the first four terms of the Taylor series if m_k is required to be a function of four moments m_0, m_1, m_2 and m_3.

In quadrature-based MOM, an infinite increase in the number of ODEs cannot always increase the accuracy when an integer moment sequence is used. The use of a fractional moment sequence was recommended to overcome this shortcoming. Our studies have confirmed that the description for the QMOM and DQMOM also applies to the TEMOM; in other words, an accurate numerical model can be constructed by introducing a fractional moment sequence. In a study of the classic TEMOM with an integer moment sequence, the classic TEMOM with a fourth-order Taylor-series expansion was observed to be less accurate than that with a third-order Taylor-series expansion and to have a narrower scope of application in terms of the geometric standard deviation (GSD) of the number distribution. This is because higher-order Taylor-series expansion leads to the appearance of higher-order moments, which might increase the numerical errors in the calculations.

To construct a closure function in terms of fractional moments rather than integer moments, we need to choose m_0, $m_{1/\phi}$, $m_{2/\phi}, \cdots, m_{\frac{H*\phi}{\phi}}$ as basis variables. H denotes the maximum number in the moment sequence to be established, whereas ϕ is the sequence number denoting the type of moment sequence to be used. The number of equations in the final moment ODEs is $H*\phi+1$. Generally, a larger number of resolved equations requires a higher computational cost, resulting in lower calculation efficiency.

We introduce a power variable, $q^k(H,\phi)$ and then expand it by using the $(H*\phi+1)$ order Taylor-series expansion as follows:

$$q^k(H,\phi) = \varepsilon_0 + \varepsilon_1 q + \varepsilon_2 q^2 + \cdots \varepsilon_{H*\phi} q^{H*\phi} + o(H*\phi+1) \quad (8.16)$$

where ε_i denotes the coefficients of the expanded polynomials, which are functions of both k and the Taylor-series expansion point q_0. Subsequently, we relate the particle volume variable v to q as follows:

$$q = v^{1/\phi}. \quad (8.17)$$

Then,

$$v^{k/\phi}(H) \approx \varepsilon_0 + \varepsilon_1 v^{\frac{1}{\phi}} + \varepsilon_2 v^{\frac{2}{\phi}} + \cdots \varepsilon_{H*\phi} v^{\frac{H*\phi}{\phi}}. \tag{8.18}$$

By incorporating (8.18) into the definition for k-th moment in Equation (8.4), we obtain a general expression for an arbitrary moment with arbitrary H and ϕ,

$$
\begin{aligned}
m_{k/\phi}(H) &= \int_0^\infty v^{\frac{k}{\phi}}(H, \phi) n(v) dv \\
&\approx \int_0^\infty \left[\varepsilon_0 + \varepsilon_1 v^{\frac{1}{\phi}} + \varepsilon_2 v^{\frac{2}{\phi}} + \cdots \varepsilon_{H*\phi} v^H \right] n(v) dv \\
&= \varepsilon_0 + \varepsilon_1 m_{\frac{1}{\phi}} + \varepsilon_2 m_{\frac{2}{\phi}} + \cdots \varepsilon_{H*\phi} m_{\frac{H*\phi}{\phi}}.
\end{aligned}
\tag{8.19}
$$

Equation (8.19) can be used to approximate any k-th moment shown in the moment ODEs.

Studies have confirmed that the closure function shown in Equation (8.19) is superior to Equation (8.15) in accuracy, especially for fractional moments at an initial evolution stage. Since 2008, several closure functions have been proposed by scientists within the framework of TEMOM for meeting specific requirements, which is shown in Table 8.1.

8.3   APPLICATION OF MOM IN THE STUDY OF VEHICLE EXHAUST PARTICULATE MATTERS

In recent years, attention has been focused on the issue of particulate emissions from vehicle engines (Chan et al., 2018). Typically, nanoparticle formation and subsequent growth in the exhaust plume is a complicated chemical/physical process and this process is extremely sensitive to dilution and mixing condition. In order to understand the fundamental science underlying nanoparticle evolution in a dilution process, it has become increasingly important that thermodynamics, fluid mechanics, aerosol kinetics and nucleation kinetics are simultaneously taken into account.

The evolution of particulate behavior in the flow field can be tracked by combining the PBE with the computational fluid dynamics. Just as discussed in Section 8.1, the high degree of the PBE makes it unsuitable to be used in the calculation. An alternative is to transfer the PBE from the particle size distribution function space $\{n(v)\}$ to the space of moments $\{m_k\}$. In this chapter, we will show an example, which implements the transformation from $\{n(v)\}$ to $\{m_k\}$ using the TEMOM, and then couple moment ODEs with Navier–Stokes equations to capture the evolution of particle under a turbulent condition.

The aim of the present example is to investigate the new particle formation by binary homogeneous nucleation in exhaust conditions and then growth by coagulation and condensation. Figure 8.3 is the Cartesian coordinate system (x, y, z) used in

TABLE 8.1

Closure Functions of TEMOM

Name	Closure Functions of TEMOM
Third(integer)-order closure function	$m_k = u_0^{k-2}\left(\dfrac{k^2-k}{2}\right)m_2 + u_0^{k-1}\left(-k^2+2k\right)m_1 + u_0^{k}\left(\dfrac{2+k^2-3k}{2}\right)m_0$
Direct TEMOM closure basis function	$\overline{m}_{nm} = \displaystyle\int_{\frac{p}{p}}^{\infty} x_1{}^n \overline{n}\left(\overline{v}_1,\, t\right)d\overline{v}_1 \int_0^{\infty} x_2{}^n \overline{n}\left(\overline{v}_2,\, t\right)d\overline{v}_2$
Generalized function(fixed maximum moment order)	$m_{k/\phi} = \displaystyle\int_0^{\infty} q^k n(v)\,dv,\ \text{with}$ $q^k = q_0^{k-2}\left(\dfrac{k^2-k}{2}\right)q^2 + q_0^{k-1}\left(-k^2+2k\right)q + q_0^{k}\left(\dfrac{2+k^2-3k}{2}\right) + \cdots$
Bi-closure basis function	$m_{kj} = m_{0j}\overline{p}^k\left(1+\dfrac{k^2}{2}-\dfrac{3k}{2}\right)+m_{1j}\overline{p}^{k-1}\left(2k-k^2\right)+m_{2j}\overline{p}^{k-2}\dfrac{k^2-k}{2}$
Generalized function (arbitrary maximum moment order)	$m_{k/\phi}(H) = \displaystyle\int_0^{\infty} v^{\frac{k}{\phi}}\left(H,\phi\right)n(v)\,dv = \varepsilon_0 + \varepsilon_1 m_{\frac{1}{\phi}} + \varepsilon_2 m_{\frac{2}{\phi}} + \cdots \varepsilon_{H*\phi} m_{\frac{H*\phi}{\phi}}$
Bivariate basis function	$m_{k,l} = \left[\left(1+kl+\dfrac{k^2-3k+l^2-3l}{2}\right)u_1^k u_2^l\right]m_{0,0}$ $+\left[\left(2k-k^2-kl\right)u_1^{k-1}u_2^l\right]m_{1,0}+\left[\left(2l-kl-l^2\right)u_1^k u_2^{l-1}\right]m_{0,1}$ $+ klu_1^{k-1}u_2^{l-1}m_{1,1}+\left[\dfrac{k^2-k}{2}u_1^{k-2}u_2^l\right]m_{2,0}+\left[\dfrac{l^2-l}{2}u_1^k u_2^{l-2}\right]m_{0,2}$

the numerical analysis. This is consistent with the experimental setup used by (Ning et al., 2005), allowing precise control of simulation validation.

In order to understand the fundamental physics of particulate evolution in the vehicle exhaust plumes, it is necessary to simultaneously investigate the evolutions of gas species phases and particles in a spatial-temporal flow field. Taking into account

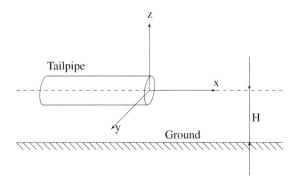

FIGURE 8.3 Schematic of the Cartesian coordinate system.

the physical terms of fluid convection, thermophoretic drift, Brownian and turbulent diffusion, Brownian coagulation, condensation and nucleation, the differential equations for the gas species and the PBE for the particle phase are:

$$\frac{\partial Y_i}{\partial t} + \frac{\partial u_j Y_i}{\partial x_j} = \frac{\partial}{\partial x_j}(D_i \frac{\partial Y_i}{\partial x_j}) + \omega_i \tag{8.20}$$

$$\frac{\partial N(v, x_i, t)}{\partial t} + \frac{\partial (u_j N(v, x_j, t))}{\partial x_j} + \frac{\partial ((u_{th})_j N(v, x_j, t))}{\partial x_j}$$

$$= \frac{\partial}{\partial x_j}\left(\Gamma \frac{\partial N(v, x_j, t)}{\partial x_j}\right) + \frac{1}{2}\int_{v^*}^{v}\beta(v - v', v')N(v - v', t)N(v', x_i, t)dv' \tag{8.21}$$

$$- N(v, t)\int_{v^*}^{\infty}\beta(v, v')N(v', x_i, t)dv' + \frac{\partial GN(v, x_i, t)}{\partial v} + J(v^*, x_i, t)\delta(v - v^*)$$

where Y_i (mol m^{-3}) is the local concentration of gas species i, D_i (m^2 s^{-1}) is the diffusivity coefficient of species i in a turbulent condition, ω_i (mol m^{-3} s^{-1}) is the source accounting for the loss of species i due to nucleation and condensation and the gain due to reaction rate and u is the velocity of the flow field. Note that $i = 1$, 2 and 3 refer to the sulfuric acid, water and carbon dioxide vapors, respectively. In Equation (8.21), N is the distribution function of particle size based on the particle volume v, G is the nucleus volume growth rate due to condensation, $\beta(v, v')$ is the coagulation kernel between particles of two volumes, J is the nucleation rate, v^* is the volume of a stable H_2SO_4-H_2O monomer, δ is the Kronecker Delta function and u_{th} is the thermophoretic velocity.

For the implementation of TEMOM, Equation (8.21) needs to be transferred to the following form (Yu et al., 2009),

$$\frac{\partial m_k}{\partial t} + \frac{\partial (u_j + (u_{th})_j)m_k}{\partial x_j} = \frac{\partial}{\partial x_j}\left(\Gamma \frac{\partial m_k}{\partial x_j}\right) + kB_1\hbar m_{k-1/3}\frac{1}{\alpha} + J(v^*)v^{*k}$$

$$+ \left[\frac{\partial m_k}{\partial t}\right]_{coa} \quad (k = 0, 1, 2) \tag{8.22}$$

where

$$\left[\frac{\partial m_0}{\partial t}\right]_{coa} = \frac{\sqrt{2}B_2(65m_2^2 m_0^{23/6} - 1210m_2 m_1^2 m_0^{17/6} - 9223m_1^4 m_0^{11/6})}{5184m_1^{23/6}}$$

$$\left[\frac{\partial m_1}{\partial t}\right]_{coa} = 0$$

$$\left[\frac{\partial m_2}{\partial t}\right]_{coa} = -\frac{\sqrt{2}B_2(701m_2^2 m_0^{11/6} - 4210m_2 m_1^2 m_0^{5/6} - 6859m_1^4 m_0^{-1/6})}{2592m_1^{11/6}}$$

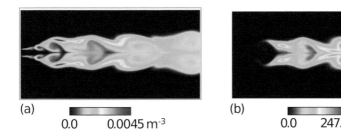

(a)
0.0   0.0045 m⁻³

(b)
0.0   247.54 m⁻³

FIGURE 8.4 Instantaneous contours of (a) total particle number concentration m_0, (b) particle mass m_1, (c) volume averaged particle diameter d_a and (d) geometric standard deviation σ_g distributions in the x-y section as the exhaust flow fully develops.

As Equation (8.22) is coupled with Navier–Stokes equations and equations for describing gas species (Equation (8.20)), the temporal evolution of a particular phase can be obtained, which is shown in Figure 8.4.

8.4 CONCLUSIONS

Micro- and nanoparticle dynamics finds applications in a wide range of powder industries, chemical engineering and environmental issues. The emphasis of the chapter is placed on the Taylor-series expansion method of moments for quantifying the evolution of particle size distribution dominated by coagulation. The numerical method can be easily applied to other mechanisms, including nucleation, condensation and breakage. The key of the method is to convert the population balance equation to moment ordinary differential equation, with it the statistical quantities of particle system can be exhibited with time. The heat and mass transfer between particle phase and continuum phase can be characterized by coupling the numerical method with the computational fluid dynamics.

REFERENCES

Balachandar, S., & Eaton, J. K. (2010). Turbulent Dispersed Multiphase Flow. *Annual Review of Fluid Mechanics, 42*(1), 111–133.

Barrett, J. C., & Jheeta, J. S. (1996). Improving the accuracy of the moments method for solving the aerosol general dynamic equation. *Journal of Aerosol Science, 27*(8), 1135–1142.

Chan, T. L., Liu, S., & Yue, Y. (2018). Nanoparticle formation and growth in turbulent flows using the bimodal TEMOM. *Powder Technology, 323*, 507–517.

Fox, R. (2012). *Quadrature-based Moment Methods for Polydisperse Multiphase Flows.* https://doi.org/10.1007/978-3-7091-1622-7

Frenklach, M. (2002). Method of moments with interpolative closure. *Chemical Engineering Science, 57*(12), 2229–2239.

Frenklach, M., & Harris, S. J. (1987). Aerosol dynamics modeling using the method of moments. *Journal of Colloid and Interface Science, 118*(1), 252–261. https://doi.org/10.1016/0021-9797(87)90454-1

Friedlander, S. K. (2000). *Smoke, dust and haze: Fundamentals of aerosol behavior.* (2nd ed.). New York, NY: John Wiley & Sons, Inc.

Fuchs, N. A. (1964). *The Mechanics of Aerosols.* New York, NY: Pagamon.

Huang, S., Vignolles, M. L., Chen, X. D., Le Loir, Y., Jan, G., Schuck, P., & Jeantet, R. (2017). Spray drying of probiotics and other food-grade bacteria: A review. *Trends in Food Science and Technology, 63*, 1–17.

Hulburt, H. M., & Katz, S. (1964). Some problems in particle technology: A statistical mechanical formulation. *Chemical Engineering Science, 19*(8), 555–574.

Keshani, S., Ramli, W., Daud, W., Nourouzi, M. M., Namvar, F., & Ghasemi, M. (2015). Spray drying: An overview on wall deposition, process and modeling. *Journal of Food Engineering, 146*, 152–162.

Lee, K. W., Chen, H., & Gieseke, J. A. (1984). Log-Normally Preserving Size Distribution for Brownian Coagulation in the Free-Molecule Regime. *Aerosol Science and Technology, 3*(1), 53–62.

Marchisio, D. L., & Fox, R. O. (2005). Solution of population balance equations using the direct quadrature method of moments. *Journal of Aerosol Science, 36*(1), 43–73.

McGraw, R. (1997). Description of aerosol dynamics by the quadrature method of moments. *Aerosol Science and Technology, 27*(2), 255–265.

Müller, H. (1928). Zur allgemeinen Theorie ser raschen Koagulation. *Fortschrittsberichte über Kolloide Und Polymere, 27*(6), 223–250.

Ning, Z., Cheung, C. S., Lu, Y., Liu, M. A., & Hung, W. T. (2005). Experimental and numerical study of the dispersion of motor vehicle pollutants under idle condition. *Atmospheric Environment, 39*(40), 7880–7893.

Otto, E., Fissan, H., & Park, S. (1999). Log-normal size distribution theory of Brownian aerosol coagulation for the entire particle size range: Part II—Analytical solution using Dahneke's coagulation kernel. *Journal of Aerosol Science, 30*(1), 17–34.

Park, S., Lee, K., Otto, E., & Fissan, H. (1999). Log-normal size distribution theory of Brownian aerosol coagulation for the entire particle size range: Part I—analytical solution using the harmonic mean coagulation. *Journal of Aerosol Science, 30*(1), 3–16.

Pratsinis, S. (1988). Simultaneous nucleation, condensation, and coagulation in aerosol reactors. *Journal of Colloid and Interface Science, 124*(2), 416–427.

Ramkrishna, D., & Singh, M. R. (2014). Population balance modeling: current status and future prospects. *Annual Review of Chemical and Biomolecular Engineering, 5*(1), 123–146. https://doi.org/10.1146/annurev-chembioeng-060713-040241

Smoluchowski, M. von. (1917). Versuch einer mathematischen Theorie der Koagulationskinetik kolloider Lösungen. *Z. Phys. Chem, 92*(9), 129–168.

Tsotsas, E., & Mujumdar, A. S. (2014). *Modern Drying Technology. Modern Drying Technology* (Vol. 1–4). https://doi.org/10.1002/9783527631728

Williams, M. M. (1985). On the modified gamma distribution for representing the size spectra of coagulating aerosol particles. *Journal of Colloid and Interface Science, 103*(2), 516–527.

Yu, M., & Lin, J. (2018). Taylor series expansion scheme applied for solving population balance equation. *Reviews in Chemical Engineering, 34*(4), 561–594.

Yu, M., Lin, J., & Chan, T. (2008). A new moment method for solving the coagulation equation for particles in Brownian motion. *Aerosol Science and Technology, 42*(9), 705–713.

Yu, M., Lin, J., & Chan, T. (2009). Numerical simulation for nucleated vehicle exhaust particulate matters via the TEMOM/LES method. *Int. J. Mod. Phys. C, 20*(3), 399–421.

Yu, M., Lin, J., Jin, H., & Jiang, Y. (2011). The verification of the Taylor-expansion moment method for the nanoparticle coagulation in the entire size regime due to Brownian motion. *Journal of Nanoparticle Research, 13*(5), 2007–2020.

Yu, M., Liu, Y., Lin, J., & Seipenbusch, M. (2015). Generalized TEMOM model for solving population balance equation. *Aerosol Science and Technology, 49*(11), 1021–1036.

Index

Note: Locators in *italics* represent figures and **bold** indicate tables in the text.